Peak Oil Prep

Peak Oil Prep

3 Things You Can Do to Prepare for Peak Oil, Climate Change and Economic Collapse

Mick Winter

Westsong Publishing
Napa, California

Printed in the United States of America.

First Edition

Library of Congress Control Number: 2006933552

ISBN-10: 0-9659000-4-5
ISBN-13: 978-0-9659000-4-1

Westsong Publishing
PO Box 2254
Napa CA 94558
www.westsongpublishing.com

Cover design: Nancy Shapiro

To my daughter Joanna, who will be designing livable, sustainable communities;

and to my wife Kathryn, who'll want to make sure that everyone in them is housed fairly.

All links in this book to websites, books, DVDs, products and other information are available at:

www.PeakOilPrep.com

The online links are checked and updated regularly to ensure their accuracy and functionality

---

Acknowledgements

Special thanks to my friend for nearly a half-century Joe Bugental for his invaluable editing skills,

and to Nancy Shapiro for once again creating an eye-catching book cover.

"Praise the Lord and pass the ammunition."

World War II song - by Frank Loesser

"Trust in Allah, but first tether your camel."

Arab proverb

Contents

Politics

Further Reading

Peak Oil

INTRODUCTION

"More than any other time in history, mankind
faces a crossroads. One path leads to despair and
utter hopelessness. The other, to total extinction.
Let us pray we have the wisdom to choose
correctly." - *Woody Allen*

By the time you're reading this, the situation in our society
may have already drastically changed. If not, it's likely to soon.
Your options will depend on what has already happened, and
what is yet to come. With *Peak Oil Prep* you have a handbook
—a how-to and where-to guide—that will help you deal with
the consequences of Peak Oil, climate change and economic
collapse, whether those consequences are minor or severe.

The book and its contents are designed so that you'll find it
useful *whenever* you come to the realization that things are
changing and not going to be the way they "always" have
been. All of the information in the book is valuable if you still
have time to prepare. Most of it is still useful *after* things have
gotten bad.

In this book you'll find suggestions and resources; ideas that
will be of help and also ideas that will spur you to think of still
other ways of dealing with, and adapting to, the new reality.

There are many factors that can lead to (or, by the time you
might read this, have *led* to) economic collapse. The financial
markets, including stocks and bonds and particularly hedge
funds. Increasing unemployment. Declining automobile sales
due to the rising cost of gasoline. The bursting of the housing
bubble. And, of course, Peak Oil.

PEAK OIL

The major immediate threat is Peak Oil. Worldwide oil production is at, or near, peak, meaning the planet—yes, even Saudi Arabia—will be producing less and less oil at a time when more and more is needed.

Oil is essential to everything on the planet—transportation, heating, agriculture, manufacturing, plastics. The planetary society demands increasing amounts every day, particularly the industrialized societies (led by the "American Way of Life is non-negotiable" United States) and the industrializing countries, such as India and China with their combined 2.3 billion people.

Peak Oil simply means that oil resources on the planet are finite and that there will come a point in time when one day less oil is being extracted than previously. And the following day even less. And so on, no matter how much exploration is done, no matter how efficient the new extraction technologies that are developed. There will come a point when less and less oil is available for the industrialized societies of the planet. Oil production will have *peaked*.

This would be alarming in itself because the needs of already-industrialized countries are dramatically *increasing*. Think growing populations. Think ever-increasing demand for power plants. Think SUVs. Think the production of more and more *stuff*.

But an even greater threat (*competition* is a better word) to available oil supplies are those previously underdeveloped countries that are now leapfrogging into the 21st century and demanding consumer equity with the long-developed—and perhaps over-developed—countries. Think one billion

Indians. Think 1.3 billion Chinese. (Even more useful, think a combined 2.3 billion Indians and Chinese with a relatively small but, by their very existence quite significant, number of nuclear weapons.)

Consumers in the West have continued to act as if petroleum resources were unlimited and indeed, in the United States at least, they have been assured by their government that resources are unlimited, thanks to the grace of God and the tax-deductible, off-shore wisdom of the oil companies.

However, the core message of Peak Oil observers is that all resources are finite, and that local communities, whether Gotham City or Hog Hollow, are too dependent on outside resources, and too little dependent on their own resources.

Peak Oil History

In 1956, M. King Hubbert, a highly-respected geophysicist in the oil industry, predicted that oil production in the lower 48 states of the United States would peak in the early 1970s. Despite his high standing in the industry, his prediction was greeted with skepticism at best, and, more commonly, with guffaws. It turned out that oil production in the United States did peak—around 1970.

In 1974, Hubbert predicted that worldwide oil production would peak around 1995. It didn't, but many believe that was only because the oil crisis and lowered production of the 1970s slowed the process down.

Scientists who support Hubbert's calculations, and who have done analyses of their own, predict that global oil production will peak no later than 2015, and possibly already did as early as 2004. The predictions vary, but all, even the non-doomsayers, recognize that oil is finite and its plentiful end will eventually come. Just as production in the United States peaked around 1970, and Alaska's North Slope did in the late 1990s, so is production peaking in the North Sea. In fact, oil production has already peaked in 50 oil-producing countries around the planet.

Only the Middle East has kept worldwide production figures from peaking. More than 70 percent of remaining oil reserves is in just five Middle Eastern countries: Iran, Iraq, Kuwait, Saudi Arabia, and Oman. (Drilling in the Arctic National Wildlife Refuge would give the United States only three months worth of oil, but the destruction would be forever.)

Peak Oil is not the same as running out of oil. It's not that there will be no more oil. It means that it's the end of cheap, plentiful oil. It means there will be ever-decreasing supplies of oil available at a time when there is ever-increasing demand.

Global oil production may not have peaked yet. Maybe. But production does seem to have plateaued in the last few years. Historically, it has been several years afterwards before it has been clear that oil production has indeed peaked in fields and regions.

However, there have been no significant new discoveries of major oil fields since the early 1960s, and discovery of smaller fields has been steadily declining. This year the countries of the world will globally consume six times as much oil as is discovered during the same year.

Interestingly, reported oil reserve capacity has risen, including a dramatic increase in the late 1980s. Skeptics point to the coincidence that those reserves increased at the same time that OPEC ruled that oil production would be limited to a percentage of known reserves. Thus: larger reserves = larger permitted oil production. Thus: six members of OPEC added 300 billion barrels of oil (with percentage increases ranging from 42% to 197%) to their reserve figures without reporting any new discoveries. Most experts agree that no scientific data have been presented to justify the claimed increases in oil reserve capacity. And then there was the Royal Dutch/Shell scandal of 2004. The company had to admit that it had knowingly overstated its reserves by more than 20 %.

Current global demand is around 80 million barrels/day, which is expected to increase by more than 2/3 by 2015. The

International Energy Agency estimated that demand worldwide would grow this year by a record 2.5 million barrels/day, up 3.2 % from last year. A third of that growth is China, whose need for oil is increasing by more than 20% a year. In 2004 China consumed 830,000 barrels/day more than than it had in 2003. (The United States, meanwhile, increased its oil demand by 14%.)

China is the second largest importer of oil in the world, followed by Japan in third place, and preceded by the United States by far and away in first place. China is banning bicycles in cities in favor of automobiles. In 2003 two million cars were sold in China, up 70% from the previous year. China may need 10 million barrels/day by 2025 (it's currently using about five million). Meanwhile, the United States used 7.2 billion barrels in 2002 (that's over 20 million barrels/day) and had to import more than half of it. It's estimated that the United States will need 50% more oil in another 20 years. Does this suggest there might be some possible disagreements between the United States and China in the future?

The countries of the world consume more than one billion barrels of oil every 11.5 days. In 2003, for the first year since the 1920s, not one megafield (500 million barrels or more) of oil was discovered. The number had been declining. There were 16 in 2000, eight in 2001, three in 2002. This is not a reassuring trend. And keep in mind that a 500-million barrel megafield is only going to provide the world with oil for about six days.

Oil companies are raking in record profits, but are not translating that into record investment in new facilities. In fact, they haven't built a new refinery in the United States since 1976. Is this just because they're greedy and want to cut expenses so they can maximize share price and corporate officer salaries? Only partially. Perhaps they also recognize there's no sense spending a lot of money expanding facilities to process stuff when the supply of that very stuff is starting to diminish.

Few people believe that current supplies of oil can keep up with the projected increase in demand. A *decreased* oil supply most certainly could not. The result would be increased demand with decreased production. Prices would rise; the cost of fuel, petroleum-based supplies (think pesticides and plastics) and transportation would increase. The cycle would become vicious, as prices increased and availability of oil and its related products decreased. Advancement of many societies would come to a halt, and already advanced societies would find themselves beginning to backslide.

Do we really need as much oil as we use? Well, for openers, all commercial pesticides are made from petroleum. Almost every internal combustion engine on the planet is powered by petroleum, and that includes more than 600 million vehicles. Each one of the tires on those vehicles takes an average of six gallons of oil to produce. Almost all transportation, whether vehicle, plane, ship or train, is powered by petroleum or by electricity generated by petroleum or natural gas (which is nearing peak itself).

All plastics are produced from petroleum. Forty percent of electricity worldwide is produced by petroleum. Americans use about three gallons a day per person. Your individual needs are probably a bit higher if you drive a 10-mile per gallon Hummer. The United States currently uses 26% of the world's oil production every day, even though it has less than 5% of the population.

Peak Oil is not a problem that has a solution. Global society runs on oil and natural gas and those fuels cannot be created out of thin air.

While advocates of alternative energy have proclaimed for decades that they have the solution—or rather replacement—with solar power, wind power, tide generators, hydro power, and the like, the reality is that our society and all its many complicated systems are powered by fossil fuel. To totally remake the global infrastructure cannot be done in many decades, let alone a few years. Even the much heralded "hydrogen economy"—proclaimed by such people as California's acting governor Arnold Schwarzenegger—holds minimal short-range hope. Hydrogen has its place, but that place is not as an energy producer. Hydrogen fuel cells *store* energy, they don't produce it. And it takes large amounts of energy to produce those fuel cells. Currently fossil fuels are the primary source of that energy.

Well, then, what are we going to do? How does our society maintain its standard of living? How do *you* maintain your standard of living? The short answer is, you won't. But there are actions you can take to minimize the damage.

All you really need to know is that oil will become more expensive, and less available. You need to find ways to use less of it, either directly as fuel for your car, or indirectly through all the products you use that are made of petroleum derivatives, including food from large corporate farms.

DryDipstick
www.drydipstick.com

For more on Peak Oil, Peak Oil Prep recommends Dry Dipstick, a metadirectory of the best Peak Oil-related websites, including news, analysis, discussion, books, videos and more.

Energy Bulletin
www.energybulletin.net

The very latest Peak Oil news.

The Oil Drum
www.theoildrum.com

Intelligent, informed discussion on Peak Oil.

You can find Peak Oil websites, books and DVDs listed at the back of this book.

CLIMATE CHANGE

Permafrost melting in Canada and Siberia, releasing still more greenhouses gases. Glaciers disappearing. Antarctic ice sheets breaking up. Hurricanes and typhoons increasing in strength and frequency. Arctic sea ice melting. The Gulf Stream "pump"--that keeps Western Europe warm—weakening and perhaps stopping (Gulf Stream Shutdown). Deadly heat waves killing tens of thousands. Dramatically decreased growing seasons worldwide. Island nations on the verge of extinction.

In the words of that great poet Bob Dylan, "Something is happening here, but you don't know what it is, do you Mister Jones?" Actually, the only people who still say nothing's wrong are outright liars or flat-earthers in total faith-based denial.

It would be nice to be able to say that global warming and climate change have peaked, that Gulf Stream Shutdown isn't going to happen (a recent study indicated a 50% chance of it happening within this century), and that things are going to settle down now. It would also be nice to say that the Easter Bunny is visiting twice this year.

Sorry, but things are not going to settle down. It's going to get far worse before it gets better—assuming it ever gets better.

The good news is that anything you do because of Peak Oil to lessen the amount of energy you use, will also help deal with climate change and global warming. You get a "two-fer".

Gulf Stream Shutdown
www.gulfstreamshutdown.com
An abrupt climate change metadirectory.

RealClimate
www.realclimate.org

Climate science from climate scientists.

Global Warming
www.ucsusa.org/global_warming
From the Union of Concerned Scientists.

Climate Crisis Coalition
www.climatecrisiscoalition.org
Individuals and groups calling for a global campaign to deal with climate change.

The Heat is Online
www.heatisonline.org
Website of Pulitzer Prize-winning journalist Ross Gelbspan.

Greenpeace
www.greenpeace.org
Activist environmental organization in 40 countries.

WorldWatch
www.worldwatch.org
Research for an environmentally sustainable and socially just society.

Climate Crash [book]
Author: John D. Cox
Abrupt climate change and what it means for our future.

The Revenge of Gaia [book]
Author: James Lovelock
By the scientist who developed the Gaia theory.

Abrupt Climate Change
www.abruptclimatechange.net
Links to relevant websites and news.

ECONOMIC COLLAPSE

The Dollar

The U.S. dollar has already peaked, as is clear by the rise in the value of the euro compared to the dollar. Two things have kept the dollar's value from being even lower. Re-investment by China and Japan into U.S. treasury bills with money from the trade imbalance (which gave the United States the money to keep buying from those countries and going further into debt), and the fact that the sale of oil worldwide is priced in dollars, or *petrodollars*.

The petrodollar, however, is not going to last. One of the little-known reasons that the United States attacked Iraq in 2003 was that Saddam Hussein had begun pricing Iraq's oil in *euros* rather than dollars. (One of the first acts of the American occupation of Iraq was to switch Iraqi oil sales back to dollars.) When the Iranians threatened to sell their oil for petroeuros (one of the main reasons the Bush/Cheney administration wanted to attack *them*)—and the Chinese and others decided they could no longer afford to prop up the dollar— the dollar started its rapid collapse. By the time you read this, you're probably living with the result. If not yet, stay tuned.

The Great Depression
www.pbs.org/fmc/timeline/edepression.htm

Public Broadcasting System (PBS).

Greatest Market Crashes of History
www.investopedia.com/features/crashes

From tulip bulb to dot.com.

Daily Life in the United States, 1920-1940 [book]
Author: David E. Kyvig

How Americans lived during the Roaring Twenties and the Great Depression.

The Great Crash of 1929 [book]
Author: John Kenneth Galbraith

The classic book on the economic collapse and Great Depression, by renowned economist Galbraith.

Hard Times [book]
Author: Studs Terkel

An oral history of the Great Depression.

Stock Market Crash
www.pbs.org/fmc/timeline/estockmktcrash.htm
PBS.

Urban Survival.com
www.urbansurvival.com

Anticipating economic collapse - long wave economic news.

Housing

One of the biggest threats to the economy is the housing bubble, particularly on the East and West coasts of the United States, and most dramatically in California, Las Vegas, and Florida, where prices have risen to obscene heights supported by cheap oil.

You might be one who believed the real estate experts when they said there was no such thing as a "housing bubble" and that your home's value was safe because "in the long run real estate *always* increases in value". (As economist John Maynard Keynes said, "In the long run, we're all dead.")

People in Texas believed that about real estate, too, in 1985. Then in 1986 the Texas housing bubble burst. Families unable to sell or avoid foreclosure simply left their house keys on the kitchen counter and walked away from their homes.

Texans won't be as surprised this time. Most everyone else will. If there's still time to sell your home, consider doing so, and then renting. If it's too late to sell, well, at least you know where you'll be living for the foreseeable future.

Food and Water

We could go on and on with the list of problems. Despite the "green revolution", people continue to starve around the world. And the green revolution is based on cheap oil and gas to produce the fertilizers and pesticides that have made such large food production possible. As energy resources deplete, so will available food.

Water supplies are dwindling, even non-potable ones. Lakes are shrinking and drying up totally; rivers, lakes, seas, oceans and aquifers are becoming polluted to the point of extreme toxicity; the level of aquifers is dropping rapidly; glaciers, whose springtime melts support hundreds of millions of lives, are disappearing.

Some Prepared, Some Didn't

While some people were sucking as much equity out of their homes as possible so they could buy an SUV, get a second home—with boat—on the lake, or take a Caribbean cruise, other people were using their money for other things—or just saving it.

Some of those people, expecting the price of gasoline to keep rising, bought hybrid or diesel cars; others changed their lives so they wouldn't need a car at all. They switched jobs, moved to towns where they wouldn't need to commute, or bought bicycles.

While some people invested in real estate, others bought gold and silver.

While some people kept extending the limit on their credit cards, others were paying off debt.

While some people bought stuff at Wal-Mart that they didn't need, and that wouldn't last more than a year before falling apart, others bought tools and materials that would last for years and help them become more self-sufficient.

While some people bought food at upscale markets, others planted their own backyard gardens, or worked with neighbors to create community gardens. Or both.

While some people made regular visits to fast-food outlets, others supported local family farms and made arrangements for ongoing delivery of fresh, healthy food.

Some people thought the idea of "peaks" was just hype, and believed it when "experts" said that "Peak Oil" was a scam perpetrated by the oil companies to raise prices. Or a scam by the Saudis to squeeze as much money as they could out of the West.

Some people believed George Bush and Arnold Schwarzenegger when they said the "hydrogen economy" would save us all. And the *Wall Street Journal* when it said that market forces would take care of everything.

Some people believed housing prices would never stop rising; the stock market would look after them (ignoring uncomfortable memories of the dot.com crash); we'd always have plenty of gas, heating oil, and electricity; and the American colossus would stride the world forever.

So Here We Are

Few people in real authority have taken (at least publicly) the problem of Peak Oil seriously. Or the federal debt. Or the trade deficit. Or the housing bubble. Or climate change.

Maybe *you* really did know better. And you were one of those few who really were concerned. But you were also one of those millions of people who were so busy working one or two or three jobs just to make ends meet that you didn't have time to prepare for anything more than the next day. It's not your fault, but you still suffer.

So, lots of things have peaked—or will shortly. The result is (or soon will be) a crashed economy; a collapsed transportation infrastructure; not enough oil, gas, electricity and energy to support our needs, let alone our desires; critical shortages of food; and huge levels of unemployment.

Not a pretty picture, is it? Oh, well; time to get on with it. You may still have some time to prepare for certain areas, but mainly you have to deal with the actual problem.

This book will help.

SCAPEGOATS

When things go badly, individuals have a tendency to blame someone. When things go badly for a country, the government and its people tend to blame someone.

When the effects of Peak Oil are truly felt and the economy collapses, the result will be a very upset populace that will likely look around for a scapegoat; some person or group on which to blame everything.

It's important to realize that no single individual or group is responsible for Peak Oil. In fact, nobody is single-handedly responsible. Peak Oil is simply a geologic reality; a result of the fact that there was always a certain amount of petroleum in the earth; no more, no less. When we as a global society have extracted half of that amount, we have reached peak, and from then on there will be less and less available.

No ethnic minorities are responsible, no racial groups, no religious groups, not even any particular political party or political system. It was all of us.

Economic collapse is the result of Peak Oil, and of many other economic factors including massive debt of both country and individuals, the more than $300 billion spent on the war in Iraq, tax cuts for the rich, hedge funds, and many other activities.

Again, no minority group, no religious group, no racial

group, nor any other traditional scapegoat is responsible.

Climate change, including global warming, is the result of our industrial society and our combustion engine-oriented transportation system. Those of us in industrialized countries who have benefited are all equally responsible for the pollution that has resulted in global warming and the changes in the global climate.

Are there some people who could have done more to prevent the current situation? Of course. We *all* could have done something, but most of us didn't.

Those who could have done the most are those politicians and other influential leaders who made decisions about national policies and denied that a problem with oil supply and global warming even existed.

There are also those people, politicians and corporations who personally profited even though they knew that society was hurt by their actions. They could have *helped* the planet and its resources rather than profited from its misuse and mistreatment. But they didn't.

Again, those who knew better and could have done more were not members of any particular ethnic, religious or racial group; nor of some other country or region of the world.

Scapegoating is of no value in dealing with the situation. There is no time for blame; only cooperation and action will help.

All links in *Peak Oil Prep* to websites and other books are also online

Peak Oil Prep
www.peakoilprep.com

For your convenience, we've gathered all the links in this book for websites, books, DVDs and other information, and put those links on one website. Once there, just click on a link to go to the source page or detailed information for each item.

> **Beyond Peak**
> www.beyondpeak.com
>
> You'll find extensive useful information at the website that was the inspiration for this book. The website has most of what is in this book, plus much more, continuously updated.
>
> Links on the above websites are checked regularly by LinkAlarm (www.linkalarm.com) to ensure that they remain active.

THE BIG THREE

Three Things to Do

> 1. Replace incandescent light bulbs with compact fluorescent light bulbs
> 2. Walk or bike instead of driving
> 3. Plant a garden

Details

With these three simple actions, you can make a major change in your energy use, personal health, and well-being. If millions of us did these actions, it would make a major change not only in our own but in our country's well-being. Two of them require little or no money and will result in benefits very quickly. The third—fluorescent bulbs—requires a small initial investment that will pay off in substantial money and energy savings over a medium period of time.

1. REPLACE INCANDESCENT LIGHT BULBS WITH COMPACT FLUORESCENT LIGHT BULBS

Incandescent bulbs are basically little heaters that also produce light. They make light by passing electricity through a small wire (*filament*). The wire heats up and glows, producing light. Unfortunately only 10% of the energy they use produces light; the other 90% produces heat. This is a very serious waste of energy.

Compact fluorescent (CF) bulbs pass electricity through a gas-charged tube. The chemical reaction produces the light. The bulb remains cooler and produces more light (measured in *lumens*) than incandescent bulbs, with less power. The average CF bulb uses 66-75% less electricity to produce the same amount of light as an incandescent bulb.

When you use less energy for your lighting, you save money. You save money because you use less electricity. Using less electricity means you're ultimately using fewer fossil fuels, such as the oil, gas or coal used in power plants. You're helping conserve those fossil fuels, and you're helping to prevent the air pollution they produce when burned.

If every U.S. household replaced just one 60-watt incandescent light bulb with a CF bulb, the pollution reduction would be equivalent to removing one million cars from the road. Think how it would be if we all replaced *all* of our bulbs.

Advantages of CF bulbs

Long Lasting - While CF bulbs cost more than incandescent, they can last anywhere from 8,000 to 15,000 hours, compared to the typical 1,000 for an incandescent bulb.

Money Saving - If you replace a 100-watt incandescent bulb with a 32-watt CF bulb (which provides the same amount of light), you can save at least $32 over the life of the bulb. Replace 10 bulbs in your home, and that's a savings of more than $300.

Cool - CF bulbs are also cool to the touch, with a temperature of less than 100° F.

Rebates - Many utility companies offer rebates or even free bulbs to homeowners. Check to see if your local company has these offers.

Selection of shades – CF bulbs are available in several shades of light, ranging from bluish to golden or bright sunlight. Pick the ones best suited for your purposes.

For inside and outside -- CF bulbs are designed for either indoor and outdoor use. Check the package to make sure you get the kind you need.

Hint

Remember "Watt Four"—CF bulbs use about ¼ of the wattage as incandescent bulbs to produce the same amount of light (lumens). So as a rule of thumb, replace an old incandescent bulb with a CF bulb of about ¼ the wattage, i.e. replace a 60-watt with a 15-watt.

---

2. WALK OR BIKE INSTEAD OF DRIVING

+ Helps with air pollution
+ Saves fossil fuels
+ Saves you money
+ Improves your health

There are so many benefits to walking that this book has room to list only a few. The physical exercise from moving your body will increase your blood flow, strengthen all the muscles in your body, loosen your joints, improve your breathing, help you lose weight over time, and increase your appetite yet help you be satisfied with lower food intake. Your mental outlook will improve and emotionally you'll feel more positive. You'll also be able to get outdoors, see your neighbors and the neighborhood, enjoy nature, reduce auto pollution, save money on gas, and feel the pleasure of being outside in the air. And it's all free.

If you absolutely can't along without driving, at least drive *less*. Here are some tips on how to do this:

+ Combine trips. Wait to combine a number of errands

into one trip
- Telecommute at least once a week
- Carpool or use public transit whenever possible
- Share errands with neighbors

Walkable Communities

It's much easier, safer, and more enjoyable to walk in a town when it's a walkable community. "Walkable" simply means that it's designed for the enjoyment—and safety—of people, not just for the unobstructed movement of cars.

> **How Can I Find and Help Build a Walkable Community?**
> **www.walkable.org/article1.htm**
>
> **Article by walkable community advocate Dan Burden.**

Is your community walkable? Here are Dan Burden's 10 keys to walkable, living communities. How well does your community rate in each category?

1. Compact, lively town center.
2. Many linkages to neighborhoods (including walkways, trails and roadways).
3. Low speed streets (in downtown and neighborhoods, 20-25 mph common).
4. Neighborhood schools and parks.
5. Public places packed with children, teenagers, older adults, and people with disabilities.
6. Convenient, safe, and easy street crossings.
7. Inspiring and well-maintained public streets.
8. Land use and transportation mutually beneficial.
9. Celebrated public space and public life.
10. Many people walking.

> **10 Keys to Walkable/Living Communities**
> www.lgc.org/freepub/land_use/articles/ten_keys/
> page01.html
>
> Full article on the 10 keys above.
>
> **Walkable Communities, Inc.**
> **www.walkable.org**
>
> Dan Burden's website dedicated to helping communities
> become more walkable and pedestrian friendly.
>
> **Center for Livable Communities**
> **www.lgc.org/center**
>
> California's Local Government Commission.

Bicycles

Bicycles are a wonderful method of transportation. For information, see page 120.

3. PLANT A GARDEN

Plant a garden in your yard (page 73) or start a community garden (page 62). Either way, you'll have a source of free, healthy vegetables and herbs, and you'll get exercise outside in the fresh air.

RELOCATION

Three Things to Do

1. Decide whether to move
2. If you want to move, research new locations
3. Move

Details

If you have no need or interest in moving to another community or country, simply skip this section.

If your timing is good, you've discovered this book before things crashed. If so, you might still have time to relocate. If you think it's a good idea, of course.

You might be fine living exactly where you are now. It's certainly easier to stay where you are than to move. No one can predict with any certainty where the best places might be to deal with the future. Many observers think that large cities are definitely not the best location. Others suggest that any town dependent on water and food that comes from a large distance may not be ideal. They suggest a small town with adequate water and nearby farms. When it comes down to it, no one knows for sure. Read the information, consider your circumstances, and make what you consider the best decision for yourself and your family.

Some believe it best to head for the hills, create a mountain fortress, and become a dyed-in-the-wool, nobody-come-near-me loner. Others believe that if a true survivalist is someone who wants to survive, the best way to do that in the 21st century is in a community.

The choice is yours. If you still have the time, take a look at the information below. You'll learn of some resources for finding a place to live in the United States and in other countries around the world.

1. DECIDE WHETHER TO MOVE

Do an analysis of where you live. Is there a good local water supply? Is there ample food produced locally? Are water and food supplies dependable? Is electricity produced nearby? Can you afford to live here if your income is reduced? Do you really want to continue living on the forty-third floor of a city high-rise? Is your community likely to be safe if there are major disruptions in day-to-day living?

Look at all the advantages and disadvantages of where you currently live in light of increasing shortages of energy and likely problems with the economy.

If you're not comfortable with the results of this analysis, ask yourself some other questions. Are your distant sources of water and energy dependable? Is there some place where you could live more cheaply? If you're concerned about sea level rising, is the area where you currently live at a desirable elevation? (Many think it is a good idea to be at least 20 feet above sea level, others suggest 50 feet).

Sea Level Rise
www.sealevelrise.net
Links to websites and news.

What about your community? Is it already Peak Oil-aware and beginning to prepare? If not, is it composed of the type of people who will face difficulties and work to overcome them? Is the leadership in your community strong enough to deal with the situation, once it recognizes the reality?

2. IF YOU WANT TO MOVE, RESEARCH NEW LOCATIONS

- ♦ Choose a town of at least 3,000 - 4,000 people and one not greater than 50,000. These are, of course, rough estimates.

It's up to you to decide if relocation is a good idea. Not many people can, or will want to, relocate, but for some it can be a sensible action. Some feel that large cities will be the best place to be, because governments will make sure those masses of voters get the food, water and other supplies that they need.

Others feel that the survivalist mode is the best way to go. Buy land off in some secluded valley, invite a few other like-minded folks, stock up on heirloom seeds, tools, lots of good toilet paper, guns and plenty of ammunition, and be prepared to look after yourselves for a long period of time.

Survivalist Info
www.survivalist.com
Emergency preparedness and disaster information, directory of survivalism websites.

The majority opinion among Peak Oil activists is that the best place to be for the future is in a small rural town. Estimates for the optimal size town run from 3,000 - 4,000 up to 50,000 or so. Too small and a town lacks the variety of skills necessary to function. Too large and it loses the ability for personal contact and community-wide consensus.

- **Look for a place with plenty of water and locally-grown food.**

Wherever you go, two of your most essential needs will be plentiful potable water and healthy food. Pick a place that has existing farms growing a wide variety of crops. Make sure that it also has additional land so that agricultural use can expand, since it's unlikely that most areas currently provide sufficient food for their existing population.

- **Choose a location at least 20 feet above sea level.**

If you're not concerned about global warming, don't worry about elevation. But if you are, consider that scientists suggest that melt of the Arctic ice cap could raise sea levels six meters (almost 20 feet) or more. Melting of the Antarctic ice sheets could raise sea levels as much as *10 times* that amount.

When could this happen? Maybe not for hundreds of years. Maybe. Then again, some scientists suggest that with the current rate of global warming, it could happen in a very few years. Unfortunately, you have to decide on your own.

If you're concerned about global warming and are moving to a coastal area, make sure your new location has sufficient altitude above sea level. You might consider at least 20 (6 meters) feet and perhaps as much as 50 feet (15 meters) above sea level.

♦ **Choose a location with sufficient sun, wind and/or moving water to produce your own alternative energy.**

Most electricity in the United States is produced in plants powered by coal or gas. Other sources are water (hydro) and nuclear. Your most reliable source is your own generated electricity. Micro-hydro (small turbines in streams and rivers) and wind are the most efficient. Solar is also excellent if you live in an area that gets sufficient sunlight.

According to National Oceanic and Atmospheric Administration (NOAA) scientist James Hansen in the *New York Review of Books*:

"During the past thirty years the lines marking the regions in which a given average temperature prevails ("isotherms") have been moving poleward at a rate of about thirty-five miles per decade. That is the size of a county in Iowa. Each decade the range of a given species is moving one row of counties northward."

Since, in North America at least, the warmer climate seems to be moving toward the north, you might want to factor this into any thoughts you have about relocation.

However, keep in mind also the comment by John Cox, author of *Climate Crash*, about the prediction that "global mean temperatures will increase two degrees Celsius in the next century. Cox says: "There's a problem with this concept. For one thing, nobody lives there. Nobody lives in the global mean. It's regional climate that matters."

Cox also points out that in the last century, the model of nature was that of "Mother Theresa", where nature changed gradually and we were all in our mother's arms, protected by a nurturing planet. Today nature is viewed as the "Dirty Harry" model—one that is "menacing and unpredictable. You don't want to mess with it."

3. Move

Waste no time. If you've decided to move, just do it, so that you can start getting organized in your new location.

Want More?

Relocating Inside the United States

If you currently live in the United States, you might want to consider simply moving to another state rather than going abroad. Things generally get much cheaper when you move away from the coasts, and the quality of life can be very good. Plus, they'll speak English and you'll usually get all the stuff you're probably used to. (Assuming *stuff* is still available.)

There are a number of excellent web sites to help you in your search for a place to move. Moving.com's city profiles provide information on hundreds of cities. The profiles include cost of living, taxes, home costs, insurance costs and quality of life factors such as population, crime, weather and education.

City Profiles
www.moving.com/Find_a_Place/Cityprofile/
Information on cities throughout the United States.

Moving.com can also help you find real estate and arrange for moving logistics. You can even compare the profiles of two cities of your choice.

Compare Cities
www.moving.com/find_a_place/compare2cities
Compare statistics on two U.S. cities.

BestPlaces
www.bestplaces.net

City Comparison
www.bestplaces.net/city
Lets you compare two cities from a list of over 3,000 places in the United States. You'll see a comparison of nearly 100 categories.

BestPlaces offers neighborhood profiles for every zip code in the United States, in-depth profiles on over 85,000 schools, a cost of living calculator that compares cities and determines what salary you'd need at a new location to maintain the same standard of living as you have now.

Plus you'll find crime rates for over 2,500 U.S. cities, most and least stressful cities, and climate profiles for 2,000 cities worldwide.

Find Your Best Place Quiz
www.bestplaces.net/fybp
Determine your own recommended best places to live.

Most Livable Communities
www.mostlivable.org
Highly rated towns in the United States.

Our favorite is Find Your Spot.

FindYourSpot.com
www.findyourspot.com

> Offers a fun quiz (it'll take you less than 10 minutes) with great questions, and it produces a list of two dozen cities that fit your quiz answers.

Find Your Spot's results for each city include an attractive downloadable four-page report with an insightful overview of the character of the area and information on climate, arts and culture, recreation, education, housing, and cost of living, crime and safety, health care, and earning a living.

You'll also find links to currently available jobs and housing, roommate services, recommended city-specific books, and travel deals if you'd like to personally visit the city.

> Cities Ranked and Rated [book]
> Author: Bert Sperling
>
> Published by BestPlaces with detailed information on over 400 metropolitan areas in the United States and Canada.

RELOCATING OUTSIDE THE UNITED STATES

Wondering where the best country is to move to? Peak Oil Prep suggests that the best way for you is to use the Web to research countries, zeroing in on those you feel most attracted to, and then visiting those countries.

People generally move to countries that are less expensive than their home country, and that have as good or better weather than where they currently live. Popular countries for Americans and Canadians are Mexico and various countries in Central America. Northern Europeans are attracted such to countries as Spain, Morocco, Turkey and Croatia.

Visit a country and spend some time there before making the actual move. Most experienced expatriates suggest living at least six months in your host country before permanently moving there. And remember, you're not going to live in an entire country, just in one specific place in a country. You don't have to love the entire country to be able to find that one special place just for you. You're looking for a region, a city, a town, even a neighborhood where you can find the qualities you want in a new home.

Here are some websites that will help in your quest for a new country.

Boomers Abroad
www.boomersabroad.com

A website devoted to giving you the best and most comprehensive information available on the Web about beautiful (and affordable) tropical coastal countries, how to get there and how to live there.

It's designed for people looking to be a retiree, a working expatriate, or simply a visitor. Boomers Abroad currently covers Costa Rica, Mexico, Nicaragua, Panama, and Uruguay (and Cuba, just in case).

Expat Communities
www.expatcommunities.com

A directory of more than 110 countries with sizeable English-speaking expatriate communities. English-language websites, organizations, online forums, meetups, local newspapers, and books of interest to current and potential expatriates.

These websites will give you, or link you to, all the information you'll need to decide if a particular country might suit your needs and deserves future exploration.

Expat Stuff
www.expatstuff.com

Wherever you end up living as an expatriate, you'll need stuff. And information. And services. This website is an excellent directory with a focus on the endless variety of information, services and tools you'll need to enjoy life and create your own income while living abroad.

You'll also find information on such things as communication, health insurance, obtaining a passport, links to country information, and general expatriate blogs and websites.

Escape Artist
www.escapeartist.com

Extensive information on expatriates and countries all over the world.

Effects of Sea Level Rise
http://flood.firetree.net/

Interactive online maps showing the effects worldwide of a sea level rise of up to 14 meters. You might want to factor this into your relocation plans.

EDUCATION

Three Things to Do

1. Break up large school districts
2. Encourage home schooling
3. Set up apprentice programs

Details

A number of Peak Oil observers believe that with declining energy resources—particularly for vehicles and heating—the days are numbered for large school districts and cross-town schools reachable only, in many towns, by school bus or private auto.

Smaller neighborhood schools are likely to return, perhaps even neighborhood schools where residents get together for group home schooling.

1. BREAK UP LARGE SCHOOL DISTRICTS

Return to the concept of neighborhood schools that kids can walk to. Chances are the concept will be forced on school districts, so they might as well start working towards it now.

Forget the yellow school buses. They can be used for their original purpose—bringing rural kids to schools—as well for providing public transportation for adults who can no longer afford to drive their own cars.

Bring control back to the people actively involved in the education of the children—the teachers during the day, and the parents in the evening.

2. ENCOURAGE "HOME" SCHOOLING, BUT COMBINE KIDS INTO BLOCK OR NEIGHBORHOOD SCHOOLS

If you're going to home school, combined home schooling will give kids the social interaction they need and want.

3. SET UP APPRENTICE PROGRAMS

Set up programs that combine study at local schools with jobs where students can learn from local businesses, professionals and craftspersons. Kids can try various occupations to see what they most like and fit, then focus on developing in-depth knowledge and skills in a chosen profession.

Want More?

Transform Neighborhood Schools Into Neighborhood Centers.

If there's room, set aside a space for retired neighbors to meet, play cards and enjoy other pastimes. In the evenings hold neighborhood and club meetings, show films, and sponsor other events. Integrate all members of the neighborhood into the school, and all students, teachers and staff into the neighborhood.

Create Neighborhood And Community Arts Programs

Create programs that provide opportunities for children to see and learn various arts, theater, music and other cultural activities.

LOCALIZE

Three Things to Do

1. "Think locally, act locally" (but communicate *globally*)
2. Buy everything, including food, from locally owned businesses
3. Never ask, or count on, higher-level organizations to do what you can do on your own at the local level

Details

Although "relocalize" and "relocalization" are the trendy words in Peak Oil circles these days, we prefer the simpler "localize".

1. "THINK LOCALLY, ACT LOCALLY" (BUT COMMUNICATE *GLOBALLY*)

The solution to almost every problem resulting from Peak Oil and economic collapse is *community*. If every community looks after itself (but not at the expense of other communities), we will have the foundation for a healthy, sustainable society. Members of a community should do everything possible to cover the needs of their own community.

However, there are excellent ideas, spectacular successes, and equally impressive failures as communities around the world strive to deal with their situations. By communicating globally—primarily on the Internet—we can all shares these ideas, successes and failures. Don't keep your community's actions a secret. Let those in other communities know what you've learned—what worked and what didn't.

2. BUY EVERYTHING, INCLUDING FOOD, FROM LOCALLY-OWNED BUSINESSES

Whenever possible, buy not just from local businesses but from local *producers* (farmers, craftsmen, manufacturers).

The more food, goods and services that are produced from within your community, the more you can depend on those things continuing to be produced. Because they will require little or no transport, they will be much less dependent on oil supplies, except for that needed in their production. Fore more on local business, see page 136).

3. NEVER ASK, OR COUNT ON, HIGHER-LEVEL ORGANIZATIONS TO DO WHAT YOU CAN DO ON YOUR OWN AT THE LOCAL LEVEL

Not only *can't* they, they likely won't even try. (Remember New Orleans and Hurricane Katrina?) Your community is the group of people who most care about your community. You and your fellow community members are also, as a collective group, those who know best the needs of your community and how best to meet those needs.

Governmental power and authority will need to devolve down to the lowest level at which appropriate decisions can be made, preferably made only by those whom the decisions affect. Those decisions can start even at the neighborhood level (see page 53), where neighbors can meet informally to determine their needs and mutually decide on the best ways to meet those needs.

Needs that can't be satisfactorily handled at the neighborhood level can be moved up to the city level. The city, since it serves a number of neighborhoods, will have specialized skills and resources not possible at the neighborhood level.

Necessary resources, skills, knowledge and experience that are not available at the city level can be obtained at the county or regional level. More and more specialized skills and resources will be available at higher levels, including state and, if desired, national.

The important principle is that decisions are made at the level at which they have their effects—and are made by the very people who will be affected.

Relocalization Network
www.relocalize.net

Support for Post Carbon Institute's Local Post Carbon Groups to increase community energy security, strengthen local economies, and dramatically improve environmental conditions and social equity.

MEDIA

Three Things to Do

1. Start a neighborhood newsletter
2. Start a community newspaper
3. Start a local radio station

Details

1. START A NEIGHBORHOOD NEWSLETTER

Keep everyone in your neighborhood informed by starting a

newsletter. You can print out copies and deliver them door-to-door, post them in central places where everyone can see them, or produce them as an email letter. Or, you can distribute newsletters using all these ways to make sure everyone has an opportunity to see them.

The newsletter can contain such information as: neighborhood projects, new tools available for sharing, neighborhood meetings, interest groups, kids' activities, classes, gardening tips. You're limited only by your resources, needs and imagination.

2. START A COMMUNITY NEWSPAPER

It would be simpler if your local newspaper(s) covered the issues of preparation for Peak Oil. If no local paper does—or will—see if *you* can write a weekly column in a paper on Peak Oil and neighborhood/community preparation. If the local paper isn't interested even in a column, consider starting your own paper, even if it's published only once or twice a month. Remember, though, that it's a huge effort and you'll likely need the help of others to make it happen.

3. START A LOCAL RADIO STATION

Currently, in the United States, it's very difficult to get permission from the federal government to start a full-power radio station. With Peak Oil, conditions may change. In the meantime, consider a low-power AM or FM station. It will probably have enough power to cover at least your neighborhood.

Prometheus Radio Project
www.prometheusradio.org

Non-profit organization supporting low power community radio stations.

Low Power FM
www.lpfm.ws

Information on legal low-power FM stations in the United States.

Low Power AM
www.lpam.info

Lots of information and a free Low Power AM Broadcasters Handbook.

If that's not workable, consider an Internet radio station, using either streaming audio or podcasting. Streaming audio lets people click on a website and listen to music, discussion, news or other content. Podcasting allows them to download the audio content to their iPod or other mp3 player in order to listen to programs whenever and wherever they wish.

Internet Radio
http://en.wikipedia.org/wiki/Internet_radio
From Wikipedia.

Starting an Internet Radio Station
www.allinternetradio.com/stationguide.asp
Step-by-step instructions.

How to Start an Internet Radio Station
http://music.lovetoknow.com/
How_To_Start_An_Internet_Radio_Station
Instructions and links to resources.

Podcasting
http://en.wikipedia.org/wiki/Podcasting
From Wikipedia.

How to Podcast
www.how-to-podcast-tutorial.com
Step-by-step guide.

MEETING TOGETHER

Three Things to Do

> 1. Organize neighborhoods
> 2. Establish community gardens
> 3. Open a Third Place

Details

Community is an abstract term and often more of a feeling than some place with rigid, geographic limits. Your community is the area that you feel most connected with, though it's probably bigger than just your neighborhood. In most cases it will be your entire town, although in a large metropolitan city, it could simply be the district in which you live.

If there's any "solution" to Peak Oil, at least on the local level, that solution is Community. The problems can be dealt with only by people working—and playing—together.

1. ORGANIZE NEIGHBORHOODS

See page 53.

2. ESTABLISH COMMUNITY GARDENS

See page 60.

3. OPEN A THIRD PLACE

Sociologist Ray Oldenburg, author of *The Great Good Place*, has coined the term "third place" for the places where community members informally hang out to socialize, discuss business deals, and talk about their lives and their community. The first place is home, the second is work. Third places are not private clubs, but locations where anyone in the community can drop by, have a cup of coffee or a beer, say hello to friends and join in on a conversation.

Third Places are disappearing everywhere as small, local businesses have a harder and harder time competing with chains (such as Starbuck's, which considers itself a third place), and as local taverns, corner stores and bookstores disappear from the scene.

A community with no third places is no community. If you don't have enough of them, start one. Start one downtown, or even in your neighborhood. In fact, every neighborhood or district in town should have at least one.

For an example, check out Third Place Commons (www.thirdplacecommons.org) in the Seattle area, which was actually created as a third place, rather than being one that evolved over the years.

Celebrating the Third Place [book]
Editor: Ray Oldenburg

Inspiring Stories About the "Great Good Places" at the Heart of Our Communities.

The Great Good Place [book]
Author: Ray Oldenburg

Cafés, Coffee Shops, Bookstores, Bars, Hair Salons, and Other Hangouts at the Heart of a Community.
Oldenburg's original book.

A Pattern Language: Towns, Buildings, Construction [book]
Author: Christopher Alexander

Architect Christopher Alexander's almost legendary opus on the elements that make a building or an entire town truly human.

Pattern Language
www.patternlanguage.com

Alexander's website is also highly recommended. You'll find lots of information and resources.

Citizens Handbook
www.vcn.bc.ca/citizens-handbook
A guide to building community. Highly recommended.

Community Collaboration
www.communitycollaboration.net
Helping communities and organizations build
collaboration and consensus.

Community Solution
www.communitysolution.org

Dedicated to the development, growth and enhancement
of small local communities.

Want More?

Intentional Communities / Ecovillages

It takes a village to raise an adult. We all need villages, even if
they're just mini-villages in urban areas. (We call those
"neighborhoods".) Many people around the planet are
working together to create new, or modify old, communities
to be sustainable, to focus on humans not cars, and to provide
healthy, people-friendly places to work, play, grow and learn.

City Comforts [book]
Author: David Sucher

How to build an urban village.

Communities Directory [book]
Editor: Fellowship of Intentional Communities

A guide to intentional communities and cooperative
living.

Creating a Life Together [book]
Author: Diana Leafe Christian

Practical Tools to Grow Ecovillages and Intentional
Communities.

Ecovillage Living [book]
Author: Hildur Jackson

Restoring the Earth and Her People.

Ecovillages [book]
Author: Jan Martin Bang

A practical guide to sustainable communities.

Global Ecovillage Network
www.ecovillages.org

A global confederation of people and communities, both urban and rural, that meet and share ideas, exchange technologies, and develop cultural and educational exchanges, directories and newsletters. An outstanding resource.

Intentional Communities
www.ic.org

A worldwide guide to ecovillages, cohousing, residential land trusts, communes, student co-ops, urban housing cooperatives and other related projects and dreams.

Toward Sustainable Communities [book]
Author: Mark Roseland

Resources for Citizens and Their Governments.

Utopian EcoVillage Network
www.uevn.org

U.S. organization dedicated to the development, growth and enhancement of small local communities with a focus on dealing with Peak Oil.

Going Local [book]
Author: Michael Shuman
Creating self-reliant communities in a global age.

Local Government Commission (LGC)
www.lgc.org

Working to build livable communities.

Center for Livable Communities
www.lgc.org/center

From the Local Government Commission. Extensive information and resources.

The Natural Step for Communities [book]
Author: Sarah James

How cities and towns can change to sustainable practices.

Partners for Livable Communities
www.livable.com

Working to improve the livability of communities by promoting quality of life, economic development, and social equity.

Stop Sprawl
www.sierraclub.org/sprawl/community
Sierra Club on livable communities. Lots of information and resources.

Community Energy Opportunity Finder
www.energyfinder.org

An interactive tool that will help you determine your community's best bets for energy solutions that benefit the local economy, the community, and the environment. Developed by Rocky Mountain Institute.

Bowling Alone
www.bowlingalone.com

Based on Robert Putnam's book on declining social capital.

Better Together
www.bettertogether.org
Tools and strategies to connect with others.

NEIGHBORHOOD

Three Things to Do

1. Organize your neighborhood
2. Carpool and share errand trips
3. Start a neighborhood garden

Details

It's good to have a home that's self-sufficient but even better to have an entire neighborhood that can cover its own basic needs. Plus your neighbors are likely to have knowledge, skills and tools that you don't have. Through mutual sharing, you're all strengthened.

The first step is to determine what your neighborhood is. In an urban area, it might be one block of a street or your apartment building. In the rural countryside, a much larger area. In a suburb, it might be several blocks, a large cul-de-sac, or a cluster of units in a condominium complex. It probably comes down to what you and your immediate neighbors feel is your neighborhood. And keep in mind that the boundaries you originally determine may change as you begin to organize.

To get through hard times, you need a support group that's bigger than just your family. That's where neighborhoods come in, or even an entire community if it's small enough.

Neighborhood cooperation allows all of you to speak in one voice to local government, whether it's city or county, making sure that your needs, interests, and opinions are respected.

You can begin to organize your neighborhood by talking to people you already know, and introducing yourself to others. Then hold a meeting at your home, or at a local school, place of worship, or other community-focused building.

Neighbors can discuss sharing tools, planting a neighborhood garden, and ensuring food and water supplies. If there are elderly and disabled in your area, you can discuss how best to look after them in any future event. Once your neighborhood starts to get organized, contact other neighborhoods and work with them to lobby elected officials, government agencies, public utilities, and the like, to make sure they're doing everything they can to assist your entire community in becoming as self-sufficient as possible.

We recommend that you also visit our Meeting Together section (page 48).

1. ORGANIZE YOUR NEIGHBORHOOD

Hold a neighborhood meeting. Make it comfortable and informal, with snacks and light beverages. Schedule regular meetings thereafter. Make at least every second or third meeting a social event as well as a planning one, so that people can socialize and get to know each other. Be sure to include teenagers and older children who wish to get involved.

+ Make and distribute a list of everyone's name, address, phone number, fax number and email address.

+ Set up a "phone tree" (a list of designated phone numbers for people to call) to notify neighbors about meetings, events and other timely information. The "phone" tree can be telephone, email, or both.

+ Identify the elderly, disabled, and others needing special care and attention.

+ Create a Skills Inventory for members of your neighborhood (medical, first aid, alternative healing, midwifery, ham radio, teaching, carpentry, electrical, sewing, plumbing, music, firearms, mechanic, welding, computer, gardening, cooking, canning, childcare, etc.). This will result from occupations, hobbies, interests and other sources. People may be surprised just how many skills they have as a group once they go through this process.

♦ Create a Resource Inventory of "things" owned by people in the neighborhood that they are willing to share. There is an incredible amount of duplication of expenses within a neighborhood. Garden equipment, specialty kitchen equipment (like canning tools), automotive repair tools, power tools, wells and other water sources, firewood, barbeques, generators, fuel, chain saws, amateur radio, and CB radios. Most of these are not items you need every day, but rather only for special occasions. Not every household needs to have a complete set of all this equipment.

♦ Keep in mind that since you now have an entire neighborhood organized, your purchasing leverage has increased. You should be able to negotiate group discounts and bulk purchases, since many of your fellow neighbors will be wanting to purchase the same things. That's just one of the many advantages of organizing your neighborhood.

♦ Use your neighborhood organization to lobby for your needs with local government. Coordinate with the city and other neighborhoods to make sure the needs of the entire community are served.

2. CARPOOL AND SHARE ERRAND TRIPS

Check with neighbors to see if you can start a carpool to work, or even just for errands around town or to a neighboring city. It's possible also to do errands for neighbors and have them return the favor for you at another time. You'll all save on time and gasoline money.

3. START A NEIGHBORHOOD GARDEN

See page 62.

Want More?

SHARE TOOLS AND EQUIPMENT

Once you've done your Resource Inventory, you can set up a system for sharing. It might be a central location with some sort of checkout procedure, or simply a list of who has what that they're willing to share.

Here are web resources with extensive information on organizing your neighborhood and bringing its members together in a stronger neighborhood community.

Center for Neighborhood Technology
www.cnt.org
Chicago-based organization showing urban communities locally and across the country how to develop more sustainably.

Neighborhood Association Manual
www.cityofsalem.net/export/departments/neighbor/manual.htm
Designed by Salem, Oregon for the city's neighborhood association chairpersons, but an excellent model for any community.

Neighborhoods Online
www.neighborhoodsonline.net
Resource center for people working to build strong communities throughout the United States. From the Institute for the Study of Civic Values and Philadelphia's LibertyNet.

Organizing for Neighborhood Development
www.tenant.net/Organize/orgdev.html
A handbook for citizen groups.

Seattle, Washington - Department of Neighborhoods
www.seattle.gov/neighborhoods
Pioneer in the United States in organizing and supporting neighborhoods.

Great Meetings

www.cityofseattle.net/neighborhoods/pubs/meets.pdf

How to prepare for a great neighborhood meeting [pdf].

Neighborhood Organizing
www.cityofseattle.net/neighborhoods/nmf/booklets/
Neighborhood Organizing word doc.pdf

Examples of several different types of successful neighborhood self-help organizations including ad-hoc project groups, a baby-sitting co-op, an "empowerment" group, and a "community building" group. Also addresses the issues of whether to hire staff, how to raise and manage money, organizational structure, membership recruitment, and organizational self-evaluation.

The Secrets of Membership Recruitment
www.cityofseattle.net/neighborhoods/pubs/secrets.pdf

Proven tips for recruiting organization members [pdf].

Superbia [book]
Author: Dan Chiras

31 ways to create sustainable neighborhoods.

SUBURBIA

Three Things to Do

1. Change all zoning to mixed use
2. Run paths through *cul de sacs*
3. Take down the fences

Details

The suburbs created in the United States—and, unfortunately, elsewhere—over the past 50 years could only have happened with widespread use of the private automobile, fueled by the cheap gasoline that resulted from cheap oil.

Few suburban tract home developments have stores, services, schools and employment within walking distance. Almost all require automobiles to get their inhabitants to and from their destinations. Those destinations are generally surrounded by vast areas of asphalt, that provide a daytime resting place for the cars that brought their humans to shop and work.

With its winding maze of streets and *cul de sacs*, the world of suburbia offers little to the pedestrian who actually wants to go somewhere, rather than simply walk around out front. Since the invention of the automatic door opener, neighbors have seen even less of each other as they drive into the garage, lower their door and enter their domicile, not to be seen until the garage doors automatically open in the morning, spewing their residents back onto the streets for yet another daily ritual.

The suburbs are a great place to be if you're a car; their value to humans is questionable, particularly if those humans are forced to go on foot because of high gas prices.

James Howard Kunstler, author of "The Geography of Nowhere - The Rise and Decline of America's Man-Made Landscape" and "The Long Emergency", calls the creation of suburbia "the greatest misallocation of resources in the history of the world".

Can suburbia be saved? The most pessimistic see its future as a wasteland, the slums of the future where empty buildings are cannibalized for their resources, and streets become impassable from lack of repair. Those who are more hopeful see suburbia transformed into a world of small villages, where the former residence-only streets are filled not only with homes—now with their own vegetable gardens—but also with shops, entertainment, small neighborhood schools, community meeting places, light industry and offices. All this with far less asphalt and much more green. In short, places where people live, work and play, all within walking distance.

But to get to this more optimistic future, many changes have to be made.

1. CHANGE ZONING TO MIXED USE

Eliminate almost all single-use zoning and change it to mixed-use zoning. Let people convert homes scattered throughout the area into mom-and-pop stores, services and restaurants. Owners will be able to live in, above, or next door, to their businesses; customers will be able to walk from their homes to the businesses.

Convert some buildings in office parks into apartments and condominiums. Turn some of the offices into businesses serving the residents of those apartments and condominiums.

Revitalize downtown areas of towns and cities by building new apartments right in the downtown area. Convert buildings, add a story of residential space above existing businesses, tear down old buildings and erect new ones for housing.

Put living units in shopping malls. Set up telecommuting centers in the malls. Use some of that vast parking area to construct new apartment buildings. Rip up some of the asphalt for gardens.

Encourage people to work from their homes, and provide tax advantages for those who do.

2. RUN PATHS THROUGH *CUL DE SACS*

Put paths and trails connecting *cul de sacs* with the *cul de sacs* and streets behind them, so that people can walk easily from one location to another without having to wander along an endless maze of sidewalks to get to what is actually a close destination.

3. TAKE DOWN THE FENCES

Take down your back and side fences and open up the area behind your home. As other neighbors do this, a beautiful parkland can be created. You'll have a private, community park for the use of residents on the block. Kids can play safely, vegetable gardens can be planted, and the feeling of community and neighborliness will be enhanced.

"N" Street Cohousing in Davis, California is a cohousing (see page 154) group that was started in 1986 when two tract homes built in the early 1950s took down their side fences. "N" Street continues to grow slowly, adding one house at a time. Currently they've expanded to 17 houses, 10 homes on "N" Street that back up to 7 homes on the adjacent street. (Two of the homes on N Street are across the street from the community but are active members.)

The removal of fences has created a beautiful open-space area that includes vegetable, flower, and water gardens; a play structure; a hot tub; a sauna; and a chicken coop, large grassy area, pond and more.

N Street Cohousing
www.nstreetcohousing.org
The Davis cohousing community's website.

Want More?

Organize Your Neighborhood

Suburbia is psychological as well as physical and geographical. Meet your neighbors. Organize events. Work and play together. Your neighborhood will be healthier, happier, and safer. (See page 53.)

Convert Lawns Into Gardens

See page 159.

Convert Schools Into Community Centers.

See page 42.

Depave

Rid the world of a small piece (at least) of concrete and let the earth breathe again. If you absolutely think you have to have something besides grass on which to park and drive your car, at least use ecological permeable pavers that still let the earth breathe and rainwater return to the aquifer.

Depaving the World
www.culturechange.org/issue10/rregister.html

How to remove asphalt or concrete from your driveway or the strip between sidewalk and curb. Turn the newly-freed space into vegetables or flowers. Then, look around town for bigger spaces.

Superbia [book]
Author: Dan Chiras

31 Ways to Create Sustainable Neighborhoods.

Suburban Nation [book]
Author: Andre Duany

The Rise of Sprawl and the Decline of the American Dream.

The Geography of Nowhere [book]
Author: James Howard Kunstler
The Rise and Decline of America's Man-Made Landscape.

The Long Emergency
Author: James Howard Kunstler

Surviving the Converging Catastrophes of the Twenty-First Century.

COMMUNITY GARDENS

Three Things to Do

> 1. Join a community garden
> 2. Start a community garden
> 3. Buy food from a community garden

Details

Community gardens, often called urban gardens, are shared plots within an urban or neighborhood setting. Gardeners share tools, knowledge and labor to produce food for themselves and others. There are an estimated 18,000 community gardens throughout Canada and the United States.

1. JOIN A COMMUNITY GARDEN

To find an existing community garden, check with your city parks department or your county or local state agricultural advisor. If that doesn't work, try the local school district. If that doesn't work, start one yourself.

2. START A COMMUNITY GARDEN

Talk to the city parks department or your county agricultural advisor for advice and a possible location. If that doesn't work, join up with some neighbors and search for a vacant lot or other suitable location.

If you find a location, contact the owner. You might have to go through the local planning department to find out who owns the lot. Tell the owner what you'd like to do. Tell him/her you're offering to clean up the lot for free. If that's not enough, promise the owner some of the produce you grow.

If you can't find a vacant lot, maybe you can team up with the neighborhood school. For more assistance, check with your local Master Gardeners group (in the United States) or any other local gardening club.

Here are some resources that will tell you exactly the steps you need to take—based on successful community garden projects elsewhere.

Here are the basic initial steps from the American Community Garden Association's document "Starting a Community Garden".

1. Form a planning committee
2. Choose a site
3. Prepare and develop the site
4. Organize the garden
5. Insurance
6. Setting up a new gardening organization

You'll find full details through their website listed below.

American Community Garden Association
www.communitygarden.org
Non-profit organization for rural and urban gardening.

Starting a Community Garden
www.communitygarden.org/starting.php
Free fact sheet from the ACGA.

Garden Your City [book]
Author: Barbara Hobens Feldt
How to start an urban garden.

Seattle P-Patch Community Gardens
www.seattle.gov/neighborhoods/ppatch
Pioneer in city-sponsored community gardening.

Seattle Tilth
www.seattletilth.org
A leader in organic community gardens.

Master Gardeners
www.ahs.org/master_gardeners

American Horticultural Society guide to Master Gardeners throughout the United States. The gardeners provide free expert advice and training to home gardeners.

3. BUY FOOD FROM A COMMUNITY GARDEN

Many community gardens donate their food to local food banks or other organizations. They may also sell at farmers markets. Check with the community gardens in your area to see where they sell their food, or if you can buy vegetables directly from them.

COMPOST

Three Things to Do

1. Start a compost pile
2. Compost all organic material
3. Use the compost in your own, or neighborhood, garden

Details

1. START A COMPOST PILE

Composting your yard trimmings and kitchen waste provides a nutrient-filled addition to the soil for your lawn and garden.

Composting - A Simple Guide To Doing It
Thanks To The *Connecticut Fund For The Environment*

+ Find a place. A corner of the yard will do. Or use a large garbage pail with holes drilled in the bottom for drainage. A wooden crate or a circular tube of chicken wire fence also works fine. Some people use trash barrels buried underground so the lid is level with the ground.

+ Start the compost pile with a six-inch layer of dead leaves, twigs, and other yard scraps. If you can, add a one-inch layer of soil.

+ Save food waste in a small, covered container in your kitchen. Include all of your vegetable and bread scraps, eggshells, fruit peels, coffee grounds, tea bags (with plastic tags or labels removed), pits, and seeds. Don't add beef, poultry, or fish leftovers because they attract pests. The smaller your scraps are, the faster they will decompose. Shred the material before adding it to the compost pile in order to speed up the composting process.

+ When the kitchen container is full, dump it onto your compost pile. To make the pile more visually attractive and to enrich the pile, you can add on top leaves, grass cuttings, or weeds pulled from your garden.

+ Continue to add your kitchen and yard scraps until your outdoor compost container is three-quarters full, or until the pile is three feet deep. Then leave that pile alone and let bacteria do their work to transform your food waste into a rich, dark, spongy soil.

+ Follow the same directions to start a second compost container or pile. By the time your second pile is complete, the contents of the first will be ready to use as a fertilizer for your yard or garden.

Tips

To work best, the compost should be damp but not soggy. If it is too wet or too dry, the composting process will slow down. Water it when the weather is dry, and cover it during heavy rains.

A well-tended compost pile has no offensive odors. If the pile begins to give off an unpleasant odor, try covering it to keep it drier.

To speed the decomposition process, you can turn the pile every few weeks during the summer with a shovel or a pitchfork. This allows oxygen to get into the compost. If steam rises when you turn the pile, don't be alarmed. This is evidence that the composting process is at work.

2. COMPOST ALL ORGANIC MATERIAL

Compost piles can include all organic yard waste and all organic scraps and leftovers from your kitchen—except meat and bones.

3. USE THE COMPOST IN YOUR OWN, OR NEIGHBORHOOD, GARDEN

Mix the compost with the soil in your garden to enrich the soil. For your own garden, see page 73. For a community garden, see page 62.

Want more?

Vermiculture

Let earthworms make your compost for you. See page 81.

Learn to Compost
www.seattletilth.org/resources/compost

Extensive information on composting from Seattle Tilth.

Compost Guide
www.compostguide.com

A complete guide to making compost.

How to Compost
www.howtocompost.org

Articles and links covering all topics about composting and organic gardening.

Cooking

Three Things to Do

1. Cook with solar energy
2. Eat more uncooked (raw and living) foods
3. Cook medicinally

Details

You probably already know how to cook. But there are ways of cooking you might not have tried that use less energy (and thus less money) and that provide greater nutrition and health.

1. Cook with solar energy

Solar cookers intensify the sun's heat to cook foods. With solar, you cook flavorful meals with free, non-polluting energy. It typically takes twice as long to cook with solar as it would with a conventional oven.

Cooking with solar can be totally free after the initial cost of building or buying a solar cooker. You can buy cookers for as little as $20 or make them yourself for a few dollars.

You can cook all year round, depending on the weather, in tropical regions and in the southern United States. Further north, you can cook whenever it is clear, except for the three coldest months of the year.

Solar Cookers International
www.solarcookers.org

Spreading solar cooking to benefit people and
environments.

Solar Cooking Archive
www.solarcooking.org

A lot of information from Solar Cookers International.
Includes free plans for making your own solar ovens.

Solar Oven Society
www.solarovens.org

Non-profit organization sells solar ovens to help support
its activities in developing countries.

Sun Oven
www.sunoven.com

Ovens for sale that are used around the world, including
large ovens used by villages.

2. EAT MORE UNCOOKED (RAW AND LIVING) FOODS

Living (such as sprouts) and raw (fruit, vegetables, seeds, nuts, and grains) foods are said to have a much higher nutrient level than foods that have been cooked. Raw foods contain enzymes, which assist in the digestion and absorption of food. These enzymes are destroyed in the cooking process.

Raw and living foods should be organic, so that you avoid the toxins found in most conventional foods. Most people agree that organic foods also taste better.

By eating raw and living foods you avoid the use of energy in meal preparation. That means less cost to you and less use of fossil fuels by society.

Living and Raw Foods
www.rawfoods.com
Frequently asked questions.

Raw Food Directory
www.buildfreedom.com/rawmain.htm
Guide to books, magazines and websites.

3. COOK MEDICINALLY

Using herbs and spices for health is more than just drinking herbal tea. Although that's an excellent way to take advantage of herbs' healing properties, you can also integrate herbs and spices into most cooking recipes for health and healing support.

Examples are garlic as an anti-microbial and to lower cholesterol, onions for colds and as an expectorant, cardomom as an antiseptic, ginger for colds and arthritis, and marjoram for indigestion.

Culpeper's Herbal
www.bibliomania.com/2/1/66
Sixteenth century herbal classic.

Spices – Exotic Flavors and Medicines
http://unitproj1.library.ucla.edu/biomed/spice
Use of spices as medicine from UCLA.

Encyclopedia of Spices
www.theepicentre.com
Information on more than 40 spices.

Growing Medicinal Herbs
www.organic-gardening.net/articles/growing-medicanal-herbs.php

Growing and using herbs for health.

HerbKits
www.herbkits.com
Indoor herb garden kits [products].

HerbKits
www.herbkits.com/medicinal.htm
Indoor medicinal herb kits [products].

Medicinal Herb Package
www.heirloomseeds.com/herb2.htm
Package of seeds from Heirloom Seeds. [product]

FOOD STORAGE

Three Things to Do

1. Dry food
2. Can food
3. Keep food cool without refrigeration

Details

Proper food storage can help food last a long time. Dried foods, such as beef jerky, dried fruits, or even dried vegetables, are popular and tasty. Canning your own food is not as common as it was fifty years ago, but is still done by millions of people. They can enjoy the taste of fresh food long after the season is over, and save money by buying bulk foods in season when they are at their lowest price.

1. Dry Food

Drying food, dehydration, is usually a simple procedure, involving little or no equipment, and will allow the food to last for extended periods of time. Foods can be dried in the sun (with or without a solar dehydrator), in an oven, or with an electric dehydrator.

Drying removes the moisture from the food so that bacteria and molds can't grow and spoil the food. The best temperature for drying food is 140°F. Higher temperatures will cook the food instead of drying it.

Fruit can be dried outdoors in the sun over a period of several days. Dehydrators, costing as little as $50 or less, speed up the drying process by lowering the humidity of the surrounding air.

Drying Foods
http://cahe.nmsu.edu/pubs/_e/e-322.html

Methods of drying foods from the College of Agriculture and Home Economics New Mexico State University.

Dehydration of Food
www.canningpantry.com/dehydration-of-food.html

Foods and how to dry them.

---

2. Can food

Canned food can include fruits, vegetables, sauces, meats, and soups. Canning food is both simple and inexpensive. If you know how to boil water, you know how to can food. You can buy a cheap canner for $20 or less, and probably get free glass jars from friends or yard sales. You'll need rings and lids for the jars, but these are very inexpensive. Some fresh food, and an hour or two of your time, and you will have done it.

Canning 101
www.backwoodshome.com/articles/clay53.html
Simple instructions for canning.

Canning Food
www.uga.edu/nchfp/questions/FAQ_canning.html
FAQ from National Center for Home Food Preservation.

3. KEEP FOOD COOL WITHOUT REFRIGERATION

You don't necessarily need a refrigerator or freezer to keep foods cool long-term. For centuries people have preserved food in cool environments such as cellars, cold water (in containers), and by using the cooling effects of evaporation, in clay pots.

The key to food storage is lowering the temperature where the food is stored. You also want to minimize exposure to light and keep it in as dry (non-humid) an area as possible. Optimum food storage prolongs shelf life, nutritional value, taste, texture and color. Date all food containers and rotate, so that you're continually selecting from the longest-stored foods.

Cool Food Storage
www.inthewake.org/b1cool.html
Various methods including water immersion, cold rooms, root cellars, ice caves, and pot-in-pot.

Pot-in-Pot Cooling
www.rolexawards.com/laureates/laureate-6-bah_abba.html
A simple technique that can be used anywhere in the world.

Refrigerator Alternatives
http://groups.yahoo.com/group/RefrigeratorAlternatives

A Yahoo discussion group on energy-efficient home refrigeration, including traditional refrigerators as well as root cellars, cooling cabinets, brine solutions and much more.

Root Cellar Basics
www.waltonfeed.com/old/cellar4.html

From Walton Feed.

National Center for Home Food Preservation
www.uga.edu/nchfp

Many types of preservation for many types of food.

Prudent Food Storage FAQ
www.survival-center.com/foodfaq

Excellent and very comprehensive information.

GARDENING

Three Things to Do

1. Start a home garden
2. Start a neighborhood/community garden
3. Start a hydroponic garden

Details

Growing your own food is essential to improving your self-sufficiency. No matter how much food you're able to grow, it all helps. The more you grow, the less you have to buy at the store.

When you grow your own food, you know exactly how it has been grown. If you wish, you can grow organically, using non-hybrid "heirloom" seeds, assuring that you get the healthiest, tastiest and most nutritious food possible.

No matter how small your living space, you can still grow some of your own food, even if it's just sprouts (see page 89), herbs (see page 189), or a couple of tomato plants.

1. START A HOME GARDEN

A garden in your yard can be as small or large as you wish and have space for. A 4'x4' area can produce a lot of food, particularly if you do "intensive" gardening.

There are many different methods, styles, techniques, and theories of gardening. We suggest you visit your local nursery and ask their advice. They know the soil and climate in your area. You can also check with your local gardening clubs or, if you live in the United States, your local agricultural advisor or Master Gardeners branch.

The basic steps of gardening are always the same: find a location, prepare the soil, plant seeds or seedlings, care for the plants, and harvest.

Here are the steps you'll follow if you use the popular Square Foot Gardening techniques recommended by Mel Bartholomew.

- Location – pick a site that has six to eight hours of sun a day
- Layout – arrange your garden in squares, not rows
- Boxes – build boxes to hold soil mix
- Aisles – space boxes three feet apart
- Soil – fill boxes with soil mix
- Grid – make a grid for the top of each box
- Care – never walk on the soil
- Select – determine a different crop for each square foot
- Plant – conserve seeds, planting only two or three seeds per hole
- Water – water by hand from a bucket

- Harvest – when each square is harvested, plant a new crop

Square Foot Gardening
www.squarefootgardening.com

"How would you like a garden filled with beautiful flowers, fresh herbs and luscious vegetables, but no weeds and no hard work?" - From the website.

Square Foot Gardening Video
Video and DVD versions of the popular book.

Acres U.S.A. Magazine
www.acresusa.com

A voice for eco-agriculture.

California Backyard Orchard, The
http://homeorchard.ucdavis.edu

If you've got the room, grow your own fruits and nuts. From the University of California.

Gardening Without Irrigation
www.gutenberg.org/etext/4512

Free downloadable e-book.

Lawns to Gardens
www.yougrowgirl.com/lawns_gardens_convert.php

How to convert your lawn to your garden.

Organic Gardening
www.organic-gardening.net

Good information and resources.

Master Gardeners
www.ahs.org/master_gardeners

American Horticultural Society guide to Master Gardeners throughout the United States. The gardeners provide free expert advice and training to home gardeners.

National Gardening Association
www.garden.org
Very extensive information.

You Grow Girl
www.yougrowgirl.com
Practical information in a friendly gardening site.

How to Grow More Vegetables [book]
Author: John Jeavons
And fruits, nuts, berries, grains and other crops than you ever thought possible on less land than you can imagine.

The Sustainable Vegetable Garden [book]
Author: John Jeavons
A backyard guide to healthy soil and higher yields.

2. START A NEIGHBORHOOD OR COMMUNITY GARDEN

See page 62.

3. START A HYDROPONIC GARDEN

Hydroponics is growing plants without soil and goes back, at least, to the Hanging Gardens of Babylon. Your hydroponic garden doesn't have to be as elaborate as the one in Babylon, but you can still produce tasty vegetables, fruits and herbs. You can also grow them inside an apartment year-round because you don't even need natural sunlight.

Studies have shown that hydroponically-grown food can be as much as 50% higher in nutrients and vitamins than field-grown food.

It's easiest if you use a kit to get started. It can cost $100 and up, but of course it's reusable again and again.

Homegrown Hydroponics
www.hydroponics.com/info

> Very comprehensive site on all aspects of hydroponics.
>
> **Hydroponics Mailing List**
> Hydroponics.org
> Mailing list for discussion about hydroponics, growing methods, tools, supplies, and more.
>
> **Simply Hydro**
> www.simplyhydro.com
> Hydroponics and organics, including the free online "Hydro U." classes.

Want More?

Garden Using Biodynamics

Biodynamics is an agriculture system initiated by philosopher/scientist Rudolf Steiner in the 1920s. More than just organic, it seeks to work with and revitalize the life forces in the plants and soil, and with the seasonal cycles of nature.

> **Biodynamic and Organic Gardening**
> www.biodynamic.net
> Excellent source of links and books.
>
> **Biodynamic Farming and Compost Preparation**
> ww2.attra.ncat.org/where.php/biodynamic.html
> Extensive information from the National Sustainable Agriculture Information Service.
>
> **Biodynamics**
> www.biodynamics.com
> Biodynamic Farming and Gardening Association.
>
> **What is Biodynamics?**
> www.biodynamics.net.au/what_is_biodynamics.htm

From Biodynamic Agriculture Australia
.
What is Biodynamics?
www.biodynamics.com/biodynamics.html
A philosophical view.

Garden Using Permaculture

Permaculture was created in the 1970s by two Australian ecologists, Bill Mollison and David Holmgren. They developed an ever-changing and expanding system of agriculture that has evolved over the years. It is a philosophy of ethics, personal responsibility and balanced ecology working with, rather than against, the natural world to create sustainable human habitats.

Permaculture farming is now done all over the world, on both large and small-scale sites.

Crystal Waters
www.crystalwaters.org.au
Permaculture village in Australia.

Gaia's Garden [book]
Author:Toby Hemenway
A guide to home-scale Permaculture.

Global Gardener
Video on Permaculture by co-founder Bill Mollison
[video].

Introduction to Permaculture [book]
Author: Bill Mollison
By the man who started the Permaculture movement.

Permaculture
http://en.wikipedia.org/wiki/Permaculture

From Wikipedia.

Permaculture [book]
Author: David Holmgren

Principles and pathways beyond sustainability by the co-founder of Permaculture.

Permaculture Institute
www.permaculture.org

Promotion and support of the sustainability of human culture and settlements.

Permaculture Resources
www.permacultureactivist.net

Good selection of articles, books and videos.

Tagari Publications
www.tagari.com

Publishes and distributes the best in Permaculture research. Established by Permaculture founder Bill Mollison.

Urban Permaculture Guild
www.urbanpermacultureguild.org

Group action to help transform urban places.

Remineralize Your Soil

Remineralization revitalizes soils by imitating natural processes and using materials ("rock dust") that are a result of glaciation, volcanic eruptions, and alluvial deposits to restore the soil with its natural nutrients. Remineralization provides slow, natural release of elements and trace minerals, rebalances soil pH, increases resistance to insects and disease, and produces larger and more nutritious crops.

You can find "rock dust" for little or no cost (at local quarries, for example), add it to your garden, and see the results for yourself.

> **On a Fad Diet of Rock Dust, How Does the Garden Grow?**
> www.gardening-guy.com/stories/storyReader$37
> Article from *New York Time.*
>
> **Remineralize the Earth**
> Remineralize.org
> Non-profit organization incorporated to disseminate ideas and practice about soil remineralization throughout the world. Site offers two free ebooks on soil mineralization.
>
> **Rock Dust Grows Extra-Big Vegetables (and Might Save Us from Global Warming)**
> www.commondreams.org/headlines05/0321-02.htm
> Article from *The Independent.*

Use Heirloom Seeds

Western society (at least in the United States) has gone from eating hundreds, if not thousands, of different varieties of vegetables, to the few, standard vegetables found at most dinner tables. The choice is limited because most of our food is grown on large, corporate-owned and monocultural farms. Such a farm would grow, for example, only one or at the most two types of tomatoes, all hybrids designed for high-yield and the ability to be tough enough to fight off disease and travel long distances. Taste and nutrition are not at the top of the corporate goals.

"Heirloom" (non-hybrid vegetables popular before the industrialization of agriculture) seed companies and organizations seek to preserve the original biodiversity, and promote sustainable, organic agriculture with traditional, vegetable, flower and herb seeds, all organic, non-hybrid, and non-genetically modified. Why? Because the foods produced are healthier and tastier. (Have you experienced the difference in taste between an organic heirloom tomato and a long-traveled supermarket tomato?) In the long-run, heirloom produce is more likely to survive and thrive in a changing environment than hybrids.

Organic Seed Alliance
www.seedalliance.org

Supporting the ethical stewardship and development of seed.

Seeds of Change
www.seedsofchange.com

Organic seeds, products and information.

Use Vermiculture

Worms are your friends. Vermiculture, also called vermicomposting or simply "worm composting", is using earthworms to break down organic matter into nutrient-rich, natural fertilizer and soil conditioner. The worms will work 24 hours a day making compost for your garden.

You can buy earthworms online or from local sources such as fishing shops where they are sold as bait. For home use, people generally use small plastic, cardboard or wood bins, which you can buy, or build or recycle yourself for a few dollars. The only other materials you'll need are soil, old newspapers and kitchen scraps.

Composting with Red Wiggler Worms
www.cityfarmer.org/wormcomp61.html

Urban agriculture notes from City Farmer.

Vermicompost
http://en.wikipedia.org/wiki/Vermiculture

Good information from Wikipedia.

Vermiculture Resources
www.empnet.com/worms/resource.htm

Good list of links.

Vermiculture Systems
www.composters.com/docs/worms.html
Buy a worm condo for your home.

Worm Digest
www.wormdigest.org
The definitive vermiculture magazine.

Worm Poop
www.wormpoop.com
Information and products.

LOCAL FOOD

Three Things to Do

1. Support local farmers
2. Start a garden (home and/or community)
3. Eat organic when possible (but *local* non-organic is better than distantly-grown organic)

Details

Food is so essential that we have to ensure that we have a steady supply of it. Growing your own is the best solution. Supporting a local farmer is also excellent, particularly since that farmer can probably turn out a lot more, and a lot more variety, than you can.

We encourage people to have both home and neighborhood gardens. Your home gardening can be as simple as herbs, sprouts, and a few tomatoes—or much more if you have the time and space. Community gardening benefits you by contact with your fellow neighbors as much as it does from the actual food itself.

What's the best way to grow and obtain food? That depends on you and where you live. We've tried to give you a wide variety of ideas here.

1. SUPPORT LOCAL FARMERS

Local farmers are a community treasure. Do everything you can to support them. Buy their produce at your local farmers' market. (If you don't have one, help start one.) Many local farms offers weekly delivery (or pickup) of food baskets. You can sign up for their service, paying monthly or quarterly. Community Supported Agriculture (CSA) can include paying for regular food baskets or actually investing in the farm.

Supermarket foods can travel an average of 1,200 miles before they reach your plate, using energy and resulting in increased air pollution. Buy food that is locally produced and in season, and you help reduce that energy requirement. Local farms are a valuable resource; your support helps to keep them alive. You also give yourself the pleasure—and health— of being able to eat very fresh, nutritious foods.

Community Supported Agriculture (CSA)

CSA farms are mutually supported by individuals and families. Through ongoing contracts, farmers deliver weekly food—usually organic—to homes. In turn, the farms receive ongoing financial support.

Local Food
http://news.bbc.co.uk/2/hi/science/nature/4312591.stm

> Greener than organic. BBC article.
>
> **Local Harvest**
> **www.localharvest.org**
> Farmers markets, family farms, CSAs, organic food.
>
> **Robyn Van En Center**
> **www.wilson.edu/wilson/asp/content.asp?id=804**
> Offers services to existing and new CSA farmers and
> shareholders, including a database of U.S. CSA farms.

2. Start a garden (home and/or community)

Whether or not you support local farms, your own garden is a must. If you have a yard, start a garden in it (replace a lawn, if necessary—see page 159). See page 73 for information on home gardens.

If you live in an apartment or condominium that doesn't have a yard, at least use containers to grow vegetables, herbs and sprouts inside your home.

Your neighborhood might want to start a larger, shared garden where many households help one another raise foods. Your community might even offer land for much larger gardens, to be shared by people from throughout the town. See more information on Community Gardens at page 62.

3. Eat organic when possible (but choose local non-organic over distantly-grown organic)

If you think the idea of organic foods is nonsense, well, then carry on. If you prefer the idea of chemical-free, pesticide-free, herbicide-free, unprocessed foods, try to get them whenever possible. Usually, they're not only chemical-free but higher in vitamins and minerals as well since the land they're grown on is almost always healthier than the soil used for mass-produced crops and livestock.

These days many of our organic foods are imported from as far away as South America or Asia. If you can buy produce from a local farmer, even if non-organic, consider if it's more worthwhile to support the efforts—and even the existence—of your local farmer than to pay for imported food that requires a large use of fossil fuels for its shipment by plane or boat.

Want More?

Slow Food

Slow Food began in Italy in 1986 as a reaction to (and protest against) the primarily American fast food industry, which has contributed not only to fast, mass-produced meals but a fast, mass-produced society in most developed countries. Slow Food's manifesto states that it is a "movement for the protection of the right to taste."

Slow Food supports and encourages quality foods and beverages, and supports local growers, chefs, winemakers and others who share their goals. Slow Food USA's mission is to "rediscover pleasure and quality in everyday life precisely by slowing down and learning to appreciate the convivial traditions of the table." It does this through local chapters called "convivia" that organize educational, cultural and, most important, gastronomic events.

Slow Foods also spawned another related movement called *Slow Cities*, which also started in Italy. Slow Cities are towns under 50,000 population that have vowed to retain their local character. They focus on environmental conservation, the promotion of sustainable development, and the improvement of urban life, encouraging both residents and visitors to slow down and enjoy life.

Slow Food USA
www.slowfoodusa.org

Slow Food activities and organizations in the United States.

Slow Cities
www.cittaslow.net

> **Official website of the Slow City movement.**

Support Local Markets

Just as it's important to support local farmers, it's important to support local retail food shops. Your local grocery store, greengrocer, butcher, dairy, cheese monger and the like are all locally-owned. They are members of the community and support local community activities, schools and groups. And they're more likely than chain stores to sell foods that come from the region where you live. In bad financial times, a large national supermarket chain can decide to close their store in your town; the owners of local grocery stores live in your community; they're not going anywhere.

Eat Less Meat

While there are health, religious, ethical, and philosophical reasons why many people don't eat meat, we suggest a very simple one. You—and society—can save money when we all cut down on meat intake. Most Americans are already getting an excess of protein. Meats can be expensive, and livestock—particularly beef cattle—consume huge quantities of feed and water in order to produce relatively small quantities of food (an estimated 2,500 gallons of water to produce one pound of beef). The land that feed stock is grown on can be better used for growing vegetables and other products.

Rainforests are being destroyed in most areas to provide land for agriculture. Most of this cleared land is being used to raise cattle. And those cattle become the fast-food hamburgers of America. Because of the large number of acres of grazing land required to raise cattle, it has been estimated that every fast-food hamburger you *don't* eat will save 55 square feet of rainforest.

The meat produced in the rainforest countries does not feed its citizens. It leaves the country. For example, the United States annually imports more than 200 million pounds of meat each year from Costa Rica, El Salvador, Guatemala, Honduras, and Panama. As a direct result, the average person in those countries eats less meat per year than the average American house cat.

Cows produce almost 20% of the methane in the atmosphere, the number two greenhouse gas. It has been claimed that a single cow is reportedly capable of emitting as much as 100 gallons of methane gas a day. Fortunately, for those of us who share the planet with cows, this gas rises into the atmosphere. Unfortunately, once it's there, the gas contributes to the warming of the planet—the "greenhouse effect".

Methane is one of the major gases contributing to the greenhouse effect. Cows are not the only producers of methane. (And apparently most of what they produce is burped—much less is emitted through their bovine posteriors.) Other ruminants such as sheep also emit methane. Termites are also reported to be a producer, as are rice paddies, marshland, and the burning of forests. But there's no doubt that the more than one billion cattle on the planet are major methane producers.

Skip that hamburger and you've saved 55 square feet of rainforest. That's 55 square feet of vegetation that will continue to be a habitat for a multitude of tropical plants and animals instead of methane-producing cows. See? The responsibility all comes back to you.

Meatless Monday
www.meatlessmonday.org

Recipes, health information and monthly promotions to encourage you to reduce your intake of meat and saturated fat.

Choose Healthy Foods

You already know what's healthy and what isn't. Sure, there are some gray areas. But there's little doubt that processed foods, and those filled with chemicals, are not as healthy as unprocessed, whole grain, naturally grown, chemical-free foods. That doesn't mean you have to be fanatic about it. Your food choices are yours, but when you're faced with the choice between a "healthy" food and a "less healthy" one, consider which is likely to improve your health.

Produce for Better Health Foundation
www.5aday.com

Provides 5-a-day recipes and tips for getting your five to nine daily servings of fruits and vegetables.

Center for Informed Choices (CIFC)
www.informedeating.org

Advocates for a diet based on whole, unprocessed, local, organically grown plant foods. CIFC believes that: placing these foods at the center of the plate is crucial for promoting public health, protecting the environment, and assuring the humane treatment of animals and food industry workers.

Eat Well Guide
www.eatwellguide.org

Free national online directory for locating producers, grocery stores, restaurants, and mail-order outlets that offer sustainable meat, including organic. The site is organized by methods of production (antibiotic free, hormone free etc.), third party certification (such as organic) and source.

World's Healthiest Foods
www.whfoods.com

Foods that are nutrient-dense, whole, familiar, readily available, affordable, and taste good.

Sustainable Table
www.sustainabletable.org

Introduces the idea of sustainability and links food purchasing choices to health, protecting the environment, rural communities and animal welfare.

Seafood Watch
www.mbayaq.org/cr/seafoodwatch.asp

The Monterey Bay Aquarium's Seafood Watch helps you
determine which fish are healthier choices.

SPECIALTY FOODS

Three Things to Do

1. Grow sprouts
2. Find mushrooms
3. Forage for wild edible plants

Details

There are many foods you can grow, or find, yourself, no
matter where you live or how large your home. They'll add
variety to your vegetable garden's produce, and save
additional money at the market.

1. GROW SPROUTS

Sprouts are an easy-to-grow, high-protein and high-fiber,
space-saving food. They're high in vitamin C and many B
vitamins, and contain enzymes that aid digestion. Everybody
should be growing sprouts.

Seeds that are frequently sprouted include alfalfa, mung
bean, soy bean, sunflower and clover, but many others can be
grown as well.

To grow them takes about 10 minutes a day, using supplies as simple as a quart glass jar, water and seeds. No matter how small your home, you're sure to have space to grow sprouts.

Growing and Using Sprouts
www.waltonfeed.com/grain/sprouts.html

Instructions from Walton Feed.

How to Grow Sprouts and Wheatgrass
www.handypantry.com/grow.htm

Instructions on jar, tray and soil methods. Also offers prouting kits and seeds, and instruction DVD/Video.

Sprouting at Home
www.cityfarmer.org/sprout86.html

Urban agriculture notes from City Farmer.

Sproutman
www.sproutman.com

Wide variety of information and products.

SproutMaster
www.sproutpeople.com

Offers information, seeds, kits and more.

---

2. FIND MUSHROOMS

Mushrooms are nutritious, delicious, and much more. Although there are more than 70,000 species of fungi, only about 250 species are considered delicious. Since there are also around 250 species that can kill you, we recommend you first start out with an experienced mushroom person. Below you'll see the website for local amateur mycology clubs that can help you.

Fungi Perfecti
www.fungi.com

Company promoting radically new environmental uses for gourmet and medicinal mushrooms. Extensive free information. Highly recommended.

MykoWeb
www.mykoweb.com

Devoted to the science of mycology (the study of the fungi) and the hobby of mushrooming (the pursuit of mushrooms). Includes the "The Fungi of California" with photographs and descriptions of over 400 species of fungi found in California (and over 1800 total photographs).

North American Mycological Societies
www.mykoweb.com/na_mycos.html

Amateur mushroom clubs in the United States and Canada.

Tom Volk's Fungi
www.tomvolkfungi.net

Excellent resources with a very useful beginner's introduction.

3. FORAGE FOR WILD PLANTS

Why shop for food when you can just pick it? Particularly when it's everywhere. Just make sure it's in an area that hasn't been sprayed or otherwise contaminated. Be certain that what you're picking to eat you have absolutely identified as safe; some deadly plants can look very similar to other safe plants. Some plants can be eaten raw; others must be cooked.

If you're a city dweller, don't despair. You might even have safe, edible food growing in a large city park. As with mushrooms, it's important to eat only those plants you can positively identify as safe. Common temperate zone plants include berries, nuts, and plants such as dandelion, asparagus, ferns and, if you live near the ocean, seaweed.

Edibility of Plants
www.wilderness-survival.net/plants-1.php

Information on what to eat—and what not.

A Field Guide to Edible Wild Plants [book]
Author: Bradford Angier

North American edible plants - where to find them, when and how to gather them, and how to prepare them.

Edible and Medicinal Plants of the West [book]
Author: Gregory L. Tilford

Includes full color photographs of every plant in the book.

Feasting Free on Wild Edibles [book]
Author: Bradford Angier

Guide to North American edible plants. Includes more than 500 recipes.

Identifying and Harvesting Edible and Medicinal Plants [book]
Author: Steve Brill

In wild (and not so wild) places.

The Illustrated Guide to Edible Wild Plants [book]
Author: Department of the Army

Produced by the U.S. Army.

The Neighborhood Forager [book]
Author: Robert K. Handerson

A guide for the wild food gourmet.

Stalking the Blue-Eyed Scallop [book]
Author: Euell Gibbons

On the immense variety of seafoods available on the ocean's edge. Hundreds of recipes.

Stalking the Wild Asparagus [book]
Author: Euell Gibbons

> The classic book on foraging for edible plants in the wild.
> Includes hundreds of recipes.
>
> The Wild Food Gourmet [book]
> Author: Anne Gardon
> Fresh and savory food from nature.

Want More?

Raise Small Animals

They're not just cute when they're young; they're food. Chickens and rabbits require little space, and can be raised in most city (and certainly suburban) backyards. Your town may have ordinances allowing raising such animals for your own use. If they don't, that could change. Perhaps even faster if you push them.

Chickens have an additional feature that rabbits don't. The hens lay eggs, usually starting at around the age of 20 weeks and continuing for as long as three years.

> Barnyard in Your Backyard [book]
> Author: Gail Damerow
>
> Beginner's guide to raising chickens, ducks, geese, rabbits, goats, sheep, and cows.
>
> Chickens in Your Backyard [book]
> Author: Rick Luttmann
>
> A beginner's guide.
>
> Beginner's Guide to Raising Chickens [video]
> www.chickenvideo.com
>
> Simple instructions for brooder preparation, unpacking mail-order chicks, treating chick health problems, moving poults into a hen yard, hen house options, flock management, how to spot common diseases and parasites and basic butchering methods.

> **Backyard Chickens**
> www.backyardchickens.com
>
> Created to help others find the information they need to raise, keep and appreciate chickens.
>
> **Raising Rabbits for Fun and Food**
> www.rudolphsrabbitranch.com
>
> Includes a primer on backyard meat rabbit raising practices.

Make Cheese

Basic cheese making is simple and economical, and the results are delicious. Eat it yourself, give it away, barter it, or even sell it.

Soft cheeses such as ricotta or cottage cheese can be made overnight. Hard cheeses such as cheddar need to age for several months. Ingredients include milk, rennet (a digestive enzyme you can buy in the supermarket), and a starter such as buttermilk or yogurt.

> **How to Make Cheese**
> www.gourmetsleuth.com/cheeserecipes.htm
>
> Recipes for making your own cheese at home - from Gourmet Sleuth.
>
> **Making Cheese at Home**
> http://schmidling.com/making.htm
>
> Basic instructions and recipes.
>
> **1...2...3...Cheese**
> www.dairygoats.com/hightor/Home%20Cheesemaking%20is%20easy.htm

Recipes and instructions from Ventura County 4H.

Fankhauser's Cheese Page
http://biology.clc.uc.edu/Fankhauser/Cheese/Cheese.html
Excellent site with instructions, photos and recipes.

DRUGS

Drugs are our friends. They help us get through times of stress. Even—and maybe especially—the "just say no to drugs" people are getting through the day thanks to drugs. Think caffeine, nicotine, ethanol and sucrose. Not to mention all those little pills doctors prescribe to keep up the spirits, calm the nerves, sharpen the mind, help relate to others, avoid aches and pains, lower the cholesterol, sleep better, wake up better, deal with allergies, and... the list does go on, doesn't it?

Some are legal, some aren't. We caution you not to do anything illegal because, well, it's illegal. We have no control over the content of the sites we link to, and of course this is all provided only for educational purposes.

We offer this information because the ability to make your own entertainment substances and mood uplifters during hard times is very useful. And you'll certainly find everyday beverages such as beer and wine to be very useful barter items —or gifts of friendship.

Three Things to Do

> 1. Make beer
> 2. Make wine
> 3. Make liqueurs

Details

1. MAKE BEER

Nothing quite quenches the thirst like a cold brew. When you make it at home, you can have real beer, as opposed to that mass-produced stuff. It's simple to make beer, though it takes some skill and experience to make *good* beer.

Brew Your Own
www.byo.com
The how-to homebrew beer magazine.

The Brewery
www.brewery.org
Total homebrewing information.

How to Brew [book]
www.howtobrew.com
Free online book.

Make Beer
www.instructables.com/id/E897F4SS6AEP28750F/
Complete instructions for making beer along with extensive step-by-step photographs.

The New Complete Joy of Homebrewing [book]
Author: Charles Papazian
Considered by many to be the bible of home brewing.

2. Make Wine

"I always cook with wine. Sometimes I even add it to the food." - *W.C. Fields*

Home winemaking has always been popular. And not everyone insists that it be Cabernet Sauvignon. Blackberry wine, dandelion wine, apple wine—there's no limit to what you can use as the basis for your creation.

A Quick Guide to Making Wine
www.homebrewmart.com/wineinst.html
28-day recipe for generic wine.

The Home Winemakers Manual
www.geocities.com/lumeisenman
Free online book.

Joy of Home Wine Making [book]
Author: Clem Stein
Even spice wines, herb wines, sparkling wines, sherries, liqueurs, and soda pop.

Making Wine from Rare Fruit
www.crfg.org/tidbits/makewine.html

Making Wines from Wild Plants
http://winemaking.jackkeller.net/plants.asp
With lots of recipes.

Roxanne's Fruit Wine Recipes
http://scorpius.spaceports.com/%7Egoodwine/fruitwinerec.htm
Everything from apple jack to watermelon wine.

Winemaking Recipes
http://winemaking.jackkeller.net/recipes.asp
More recipes than you'll be able to make in several lifetimes.

> **Wine World FDW**
> www.wineworldfdw.com
>
> An excellent site on making wine from a whole bunch of different things, none of them grapes. Highly recommended.

3. Make Liqueurs

Man does not live by beer and wine alone. There are also liqueurs, originally crafted centuries ago by monks as healing elixirs. At least that's what they claimed.

Liqueurs, also called cordials, are sweet alcoholic beverages made from fruits, herbs, barks, seeds, spices, and flowers that are usually delicious and frequently even healthy.

Many liqueurs are best when aged for some months, but some can be ready in a few days, if you're really impatient.

> **The Alaskan Bootlegger's Bible [book]**
> Author: Leon W. Kania
>
> How to make beer, wine, liqueurs and moonshine whiskey.
>
> **The Art of Making Wine and Liqueurs**
> Author: B. Sampson
>
> Step-by-step guide to home wine and liqueur making— from flower, fruit, and sparkling wines to sloe gin and cherry brandy. 100 recipes.
>
> **Classic Liqueurs [book]**
> Author: Cheryl Long
>
> The art of making and cooking with liqueurs.
>
> **Cordials From Your Kitchen [book]**
> Author: Pattie Vargas
>
> Easy, elegant liqueurs you can make and give. Recipes for fruit, nut, spice, coffee, and cream liqueurs, plus flavored

brandies, rums, and vodkas.

Liqueur
http://en.wikipedia.org/wiki/Liqueur
From Wikipedia.

List of Liquers
http://en.wikipedia.org/wiki/List_of_liqueurs
Information on more than 75 commercial liqueurs.

Liqueur Making
www.guntheranderson.com/liqueurs.htm
Principles and techniques.

Making Liqueurs and Cordials
www.liqueurweb.com
Excellent directory of resources.

Recipe Links
www.liqueurweb.com/links.htm
Recipe sites.

Want More?

ALCOHOL

Humanity's oldest and favorite drug. Delightful in so many different ways. Every culture has its own favorites. In some countries, such as the United States, it is illegal to distill alcohol without a license, and you probably can't get a license.

Building a World Class Home Distillation Apparatus
www.moonshine-still.com
A step-by-step guide to building a relatively sophisticated home distillation apparatus that produces a highly refined distillate. The still is made from commonly available materials, with simple hand tools, and can be built for under $US 100.

"The purpose of the website is to educate and inform those of you who are interested in this subject. It is not to be construed in any fashion as an encouragement to break the law.

"If you believe that laws against home distillation of alcohol are incorrect, you are urged to contact your representatives in government, cast your vote at the polls, write newsletters to the media, and in general, try to make the changes in a legal and democratic manner."

Gert Strand
www.partyman.se

Liquor and liqueur essences, candy shots, turbo yeasts and supplies for distilleries, home distillers and home brewers.

Home Distillation of Alcohol
www.homedistiller.org

Everything you'd want, or need, to know about home distillation. Lots of recipes. Highly recommended.

Home Distillation Handbook
www.home-distillation.com/onlineorder.htm

Purchase downloadable book.

COFFEE

According to the Whatcom Seed Company: *Coffea arabica* is easy to grow indoors, makes a very attractive houseplant, and if it likes you well enough it will even reward you with flowers and berries. A six-foot plant can produce two to four pounds of coffee a year. Grow in medium light, or filtered or indirect sunlight. Use a rich, acid soil kept moderately moist. Peat moss in the potting mix will help provide acid conditions. Ideal temperatures are between 60° F and 85° F. Give the roots room to grow. Hardy to 28° F.

> **Coffee Arabica**
> http://seedrack.com/02.html
> Coffee seeds from Whatcom Seed Company.
>
> **Coffee Growing at Home**
> www.coffeeresearch.org/coffee/homegrowing.htm
> Step-by-step instructions.
>
> **Coffee Substitutes**
> www.coffeeresearch.org/coffee/homegrowing.htm
> From Civil War days.
>
> **Growing Coffee Arabica at Home from Seed**
> www.sweetmarias.com/growingcoffeeathome.html
> Good instructions and photographs, but the author isn't very optimistic about success.
>
> **Grow Your Own Herbal Tea**
> www.hgtv.com/hgtv/gl_herbs/article/
> 0,1785,HGTV_3595_2045629,00.html
> Information on growing common herbs for tea.

MARIJUANA

It's not currently legal in the United States, but if things change, you might want to read this information:

> **Marijuana**
> www.beyondpeak.com/drugs-
> beyondpeak.html#marijuana
> All the information you'll need.

TOBACCO

Yes, we know it's unhealthy. Or at least the mass-produced commercial products are. But what if it's organic, without pesticides and all those chemicals? Tobacco has medicinal value, makes an extremely valuable ornamental plant and flower garden specimen, and is used to make one of nature's finest biodegradable, all natural pesticides.

Coffin Nails
www.coffinnails.com

Grow and smoke your own tobacco.

Growing and Processing Tobacco at Home [book]
Author: Jim Johnson
www.seedman.com/Tobacco.htm#1

A guide for gardeners. Tax-free and chemical-free tobacco. And it's legal.

Tobacco Plant Information
www.boldweb.com/greenweb/nicoinfo.htm

Planting and raise, curing, other non-smoking uses, and U.S. policy.

Tobacco Plant Seeds
www.boldweb.com/greenweb/tobacco.htm

Enough seeds for 25 or more plants.

AIR

Three Things to Do

1. Purify air with houseplants
2. Plant trees
3. Drive less

Details

Ideally you're living in a location that has lots of fresh air. However, many people aren't able to live in such a place. You may live or work in a building with no circulation, and filtered recycled air. Or you may live in a city with bad pollution, from automobiles, factories or other causes.

There are still things you can do. Probably the most important is to keep your living and working space as healthy as possible. The best and easiest way to do this is with plants and trees.

1. PURIFY AIR WITH HOUSEPLANTS

B.C. Wolverton, an environmental consultant and retired NASA scientist, describes in his book "How to Grow Fresh Air" how houseplants filter toxins from the air in your home. Plants break down the chemicals commonly found in the home, including those from paint, plastics, cleaning supplies, and synthetic carpets.

Plants such as Boston ferns and lady palms are among the best and are easy to care for. The ferns remove formaldehyde and the palms are excellent for removing ammonia. Spider plants deal with benzene, which is found in house paint. Plants are also excellent air purifiers in offices, which are frequently closed systems with no natural ventilation.

How to Grow Fresh Air [book]
Author: B.C. Wolverton
50 house plants that purify your home or office.

---

2. PLANT TREES

Trees absorb carbon dioxide and produce oxygen. They also lower air temperature, provide shade and shelter, cut down noise pollution, improve water quality, and stabilize soil. That's a lot more than we humans do. The least we can do is plant more trees.

A single tree can remove as much as 25 pounds of carbon dioxide a year. An acre of forest can absorb up to 5 tons a year. This country's forests—which are decreasing at nearly one million acres a year—can remove more than 1.5 billion tons of carbon dioxide a year.

If one out of every two Americans planted one tree by their home or business, we could save $4 billion in energy costs and reduce carbon dioxide emissions by 18 million tons a year. The shade from three trees around your home could reduce your air-conditioning needs up to 50%.

According to the U.S. Department of Agriculture, the net cooling effect of a young, healthy tree is equivalent to 10 room-size air conditioners operating 20 hours a day. The U.S. Forest Service says that trees properly placed around your home can reduce air conditioning needs by 30% and save 20-50% in heating costs. Their attractiveness also increases the market value of your home, and being around them relieves stress.

3. DRIVE LESS

Drive your car as little as possible. Carpool, walk, bicycle, use public transit—anything to reduce the amount of carbon dioxide released by the burning of fossil fuels.

HEALTH

Three Things to Do

1. Walk 30 minutes a day
2. Get more sleep
3. Eat better

Details

It's a cliché to say that without health you have nothing, but it's still obviously true. And the healthier you are, the better your life is likely to be. If you take responsibility for your own health, you can bring about remarkable results. The tips we offer here can bring about a dramatic improvement in how you feel, yet they cost nothing.

1. WALK 30 MINUTES A DAY

See page 30.

2. GET MORE SLEEP

More than 70% of Americans of all ages suffer from sleep deprivation. One of the best things you can do is get more sleep, and a more regular sleep. Based on what time you have to get up in the morning, set a bedtime at least seven hours prior to the time you rise. (After a while, try to extend this until you're getting eight hours of sleep at night.)

The amount of sleep needed depends on the individual, so experiment. Studies show that most Americans think they're getting enough sleep, and those same studies also show that they aren't.

Having trouble sleeping? Visit www.insomniatips.net for a number of ideas on how to deal with insomnia and get to sleep. Here are just a few of the 41 free tips.

- **Drink Warm Milk** - A glass of warm milk 15 minutes before going to bed will soothe your nervous system. Milk contains calcium, which works directly on jagged nerves to make them (and you) relax.

- **If You Can't Sleep, Get Up** - Don't lie awake trying to get to sleep any longer than 30 minutes. If you try for that long, get up. Do something quiet and non-stimulating. When you feel tired again, go back to bed.

- **Visualize Something Peaceful** - Just lie there with your eyes closed, and imagine you're in your very favorite, most peaceful place. It may be on a sunny beach, swinging in a hammock in the mountains or your back yard, or all alone in a cave in the Himalayas.

  Wherever it is, imagine you are there. You can see your surroundings, hear the peaceful sounds, smell the fragrance of the flowers, and feel the warmth of the sun or whatever sensations are there. Just relax and enjoy it—and drift off to sleep.

  Once you've found a place that's especially peaceful and effective, you'll find that the more you use it, the more you can count on it to help you relax and get to sleep. Its comfort and familiarity will make it more and more effective.

- **Visualize Something Boring** - The beauty of this one is you can turn a negative into a positive. Just visualize that you are someplace that you have always found extremely boring. It could be listening to a particular teacher who was so boring that he or she almost always put you to sleep. Perhaps it's some friend or acquaintance whose incessant talk and theories put you to sleep. Maybe it's your work, or maybe it's your commute each day.

  Whatever it is, visualize it. And recapture that bored, tired, heavy, sleepy feeling that you always experience. Let that feeling spread through your mind and all through your body till you're filled with complete tiredness and sleepiness.

> **41 Simple Tips to Help You Get to Sleep**
> www.insomniatips.net
>
> Free home and folk remedies to deal with insomnia and help you sleep.

3. EAT BETTER

We don't have to tell you what you should be eating more of, and what you should be eating less of. You already know. But don't try to change all at once. That way lies failure.

Tomorrow add to your daily diet just *one* thing that you know you should be eating but haven't been eating, either because you don't really care for it or because you "just don't get around to it". Next day add another food. Keep doing that each day, experimenting with a different food.

After one week, eliminate one food each day that you know you should really eat less of, or not at all. Do that as well for one week, again with a different food each day.

When we suggest this, we don't mean keep adding (or subtracting) an additional food each day. We mean just experiment, one a day less, or one a day more. After you've done this for a couple of weeks, see how you feel. Then see which foods you can, and want to, add—or remove— permanently.

Want More?

Use Alternative Medicine

Pharmaceutical drugs are already extremely expensive. After an economic collapse, they could be not only expensive but hard to get. Now is a good time to learn about alternative ways of healing and staying healthy. (See page 186).

Practice Breathing Deeply

Most of us breath lightly and very shallowly, using only the top part of our lungs. The result is a deterioration of our lung capacity, and less oxygen in our blood system. By simply retraining your body to breathe more deeply, you'll notice a dramatic improvement in how you feel and how you think. (See page 195.)

Living to 100 Life Expectancy Calculator
www.agingresearch.org/calculator/

See where your present lifestyle might get you.

Cooperative Health Insurance
www.healthdemocracy.org

An interesting proposal to establish cooperative health coverage in the United States.

Where There Is No Dentist [book]
Author: Murray Dickson

A community dental care manual designed to help people care for their teeth and gums.

Where There Is No Doctor [book]
Author: David Werner

A community health care manual. Vital information on diagnosing and treating common medical problems and diseases, giving special emphasis to prevention. Includes sections detailing effective examination techniques, home cures, correct usage of medicines and their precautions, nutrition, caring for children, ailments of older individuals and first aid. Translated into more than 90 languages.

WATER

Three Things to Do

1. Use less water
2. Capture rainwater
3. Purify your own water

Details

1. USE LESS WATER

The average person in the United States uses between 100 and 250 gallons of water a day. It's possible—and you may have no choice—to get by on a lot less.

Showers account for 2/3 of the average water heating cost, and 20% of water usage. Take shorter showers and you'll save money and energy. You should at least have low-flow shower heads. Also try turning the shower off while soaping up, then back on to rinse.

Your toilet should be low-flow. Ideally, it would be a composting toilet using no water at all. Waterless urinals are also available, although more often seen in commercial establishments.

More advanced, but effective, water-saving techniques include using graywater (from sinks, showers and washing machines) for watering gardens, and *xeriscaping*—using local, drought-tolerant plants that may need no watering at all.

H2ouse Water Saver Home
www.h2ouse.org

A guide to conserving water from the California Urban Water Conservation Council.

Low-flush Or Compost Toilets

Toilet tanks should use no more than 1.6 gallons per flush. Toilets are the largest water-wasters in the home, wasting gallons of water with every flush. If you can't put in a low-flush toilet, at least add plastic displacement bags to your toilets. (Don't use bricks. They just break apart and cause problems.)

Stop Leaks

Leaks can waste many gallons of water a day. Check for toilet leaks, as well as leaky faucets and water pipes. For toilets, put a drop or two of food coloring in your tank. If the color appears in the bowl, you've got a leak.

Xeriscaping

Xeriscaping means using native plants that are at home with your local climate conditions. They're natural, and they need no watering.

Use Less Water and Save

Here are some normal household uses of water, how much water they use normally, and how much less they use if you try to conserve.

Shower – Normal, 25 gallons – Wet down, soap up, rinse off, 4 gallons.

Brushing teeth – Tap running, 2 gallons – Wet brush, rinse briefly, one-quarter gallon

Shaving – Tap running, 10 gallons – Fill basin, 1 gallon.

Dishwashing – Tap running, 15 gallons – Wash in dishpan or sink, 5 gallons.

Automatic dishwasher – Full cycle, 16 gallons – Short cycle, 7 gallons.

Washing hands – Tap running, 2 gallons – Fill basin, 1 gallon.

Flushing toilet – Normal, 3.5-5 gallons – Low-flow toilet, 1.6 gallons.

Washing machine – Full cycle, top water level, 40 gallons. Short cycle, minimal water level, 25 gallons.

2. CAPTURE RAINWATER

Capturing rainwater is called "rainwater harvesting". It basically means water delivered directly to your home from the skies at no charge. You just have to catch it, clean it, and store it. Assuming you have enough rainfall in your area, it's a way to become self-sufficient—or to at least supplement your municipally-supplied water.

Rail Barrel Guide
www.rainbarrelguide.com
How to use rain barrels for water collection. Excellent overview and specifics.

Raincatcher
www.raincatcher.org
Harvesting natural rainwater to quench the world's thirst.

Garden Watersaver
www.gardenwatersaver.com
System uses any container for rainwater collection. [product]

Natural Rain Water
www.naturalrainwater.com
Information, products, and how to make a rain barrel.

Rain Barrel Guide
www.rainbarrelguide.com
How to use rain barrels for water collection.

3. Purify your own water

If, unlike hundreds of millions of people around the world, you're lucky enough to have access to water, you still need to make sure it's safe.

There are a number of devices and chemicals you can buy to purify water, ranging from hand-held filters to iodine tablets or hydrogen peroxide.

Here's an incredibly simple—and free—way of disinfecting water using solar radiation and plastic bottles.

Solar Water Disinfection (SODIS)

The Solar Water Disinfection (SODIS) process is a simple technology used to improve the microbiological quality of drinking water. SODIS uses solar radiation to destroy pathogenic microorganisms which cause waterborne diseases.

SODIS is ideal for treating small quantities of water. Contaminated water is filled into transparent plastic bottles (PET bottles are best) and exposed to full sunlight for six hours.

1. Wash the bottle well the first time you use it.
2. Fill the bottle to the top and close the lid.
3. Place the bottle on a corrugated iron sheet or on the roof.
4. Expose the bottle to the sun from morning until evening for at least six hours.
5. The water is now ready to drink.

Solar Water Disinfection Process
www.sodis.ch
For more detailed information.

Colloidal Silver

A centuries-old technique for purifying water is silver, used frequently these days in high-tech air and water purification systems, such as recirculating air on aircraft. For more information, see page 188.

Blue Future Filters
www.bluefuturefilters.com

Low-tech slow sand filters [products].

Ceramic Water Filter
www.potpaz.org/pfpfilters.htm

Low-tech colloidal silver-enhanced ceramic filtration and purification of water from Potters for Peace.

Silver Ceramic Water Purifiers
www.purifier.com.np

Low-cost, high-performance water purification.

Solar Water Disinfection Process
www.sodis.ch

An incredibly simple—and free—way of disinfecting water using solar radiation and plastic bottles. This information needs to be spread all over the world.

Treatment of Water to Make it Safe for Drinking
www.cdc.gov/travel/water_treatment.htm

From the Centers for Disease Control.

Water Purification
http://en.wikipedia.org/wiki/Water_purification

From Wikipedia. Includes overview of water treatment methods, other purification techniques, and portable water purification.

Water Purification for the Traveler
www.artoftravel.com/10water.htm

Obtaining safe water, filters vs. purifiers, selection criteria, comparison of filters.

Water Treatment FAQ
www.1stconnect.com/anozira/SiteTops/water/waterFAQ.htm

Storage and purification.

WaterCure.com
www.watercure.com

Website of the author of "Your Body's Many Cries for Water".

The Wonders of Water
www.watercure.com/wow/wonders_of_water.html

Healing uses of water.

Your Body's Many Cries for Water [book]
Author: Fereydoon Batmanghelidj

You are not sick; you are thirsty.

Bottled Water: Pure Drink or Pure Hype?
www.nrdc.org/water/drinking/nbw.asp

Research results on high cost and health hazards of bottled water.

CARS

Three Things to Do

1. Buy a fuel-efficient car
2. Get the best fuel efficiency you can with your existing car
3. Carpool

Details

Most environmentalists and others concerned about fossil fuels and oil depletion are focused on "environmentally responsible" ways of fueling automobiles. In other words, their goal is to keep everyone driving cars but to do it in a more efficient and "green" manner.

Some see the issue differently, and realize that one of our core problems is our society's dependency on the automobile. They say that what is needed is a dramatic change in our transportation systems and our design of cities, towns and especially our suburbs. They also note the huge amount of energy and resources needed for the manufacture of automobiles, and the equally huge amount of land area (much of it formerly agricultural land) required for highways, roads and parking.

That said, we will still suggest ways you can make your use of automobiles "better". But we urge you to find ways to lessen, or even avoid, the use of *any* type of automobile.

In fact, selling your car and using all other forms of transportation (foot, bike, bus, train) could be the best move you could make. You'll stop polluting the atmosphere, you'll save large amounts of money (gas, repairs and maintenance, insurance), you'll probably get more exercise, and you'll be able to enjoy life at a slower-pace, paying more attention to your surroundings and its inhabitants.

If you absolutely need your own car to get to work, why not change jobs? Or move to a location that no longer requires a car to get to work? A car is a major burden; you'll be surprised how liberating it can be to get rid of it.

1. BUY A FUEL-EFFICIENT CAR

Go for the most fuel-efficient car you can find. Currently, hybrids (electric/gas) are at the top of the fuel-efficiency charts. We suggest the likelihood that a hybrid or other high-mileage car will have a high resale value; in fact, once people start realizing how high gas prices can go, your hybrid could appreciate in value far beyond what you paid for it.

Hybrid Cards
www.hybridcars.com

Comprehensive information and links.

Zap SmartCar
www.zapworld.com/cars/smartCar.asp

Not a hybrid, pure gas-driven. But great mileage.

---

2. GET THE BEST FUEL EFFICIENCY YOU CAN WITH YOUR EXISTING CAR

Use your car's air conditioner as little as possible. Air conditioners can decrease your mileage as much as 20%. They're more efficient on the highway, much less efficient in stop-and-go city driving.

Drive at a steady speed. Speeding up and slowing down greatly increases fuel usage. Avoid jack rabbit starts, and plan ahead to slow down at traffic lights and stop signs.

Drive at moderate speeds. Most cars get their best mileage somewhere around 50-55 miles per hour. Going faster increases wind resistance and decreases fuel economy by as much as 6% for every five miles an hour over your optimal speed.

Keep your car well maintained. Keep your car in the best possible condition with regular tune-ups, and you'll get the best possible mileage. A badly maintained engine can use as much as 50% more fuel and produce 50% more pollution.

When you see a hill coming up, start accelerating before you get to it. It takes much more fuel to accelerate going up a hill than on a level road.

It's better to turn off your car than to let it idle for a long period of time. More than 10 seconds of idling uses more gas than you'll use to restart.

Most cars these days don't need a long warm-up time. Start driving slowly as soon as you can. The car will warm up along the way.

Measure your tire pressure monthly. Keep the tires at the properly level and you'll get better gas mileage and a smoother ride. You can improve your gas mileage by around 3.3 percent by keeping your tires inflated to the proper pressure. According to the U.S. Department of Energy, under-inflated tires can lower gas mileage by 0.4 percent for every one pound per square inch drop in pressure for all four tires.

Replacing a clogged air filter can improve your car's gas mileage by as much as 10 percent. Your car's air filter keeps impurities from damaging the inside of your engine. Not only will replacing a dirty air filter save gas, it will protect your engine.

3. CARPOOL

Carpool to work, or even on errands, whenever possible. Talk to co-workers about sharing driving and see if your company has a policy of subsidizing carpooling, since it requires fewer parking spaces in the company parking lot.

Most municipalities and transit agencies can give you information about carpool networks, where you can find people going to the same area in which you work.

---

eRideShare.com
www.erideshare.com

A free service for connecting commuters or travelers going the same way.

Want More?

Car Sharing

Some communities now offer *car sharing*, where you have use of a car whenever you need it, but you don't have to maintain it.

With car sharing, you pay for a car, van or truck only when

you use it. Cars are available 24 hours a day, and you can reserve by phone or Internet. You never pay for repairs, insurance or monthly parking.

CarSharing Network
www.carsharing.net
Find out which cities around the world have car sharing.

Car Sharing
http://en.wikipedia.org/wiki/Car_sharing
From Wikipedia.

CarSharing.us
http://carsharing.us
Blog hosted by car sharing pioneer Dave Brook.

What is Carsharing?
http://ecoplan.org/carshare/general/basics.htm
Carsharing and its benefits.

World Carshare Consortium
www.worldcarshare.com
Worldwide news and links, and online forum.

Use Biodiesel

Biodiesel is a processed fuel for diesel engines derived from biological sources, usually rapeseed or soybean oil. Because biodiesel is much cheaper than gasoline, many people are now making their own biodiesel, or at least buying it, for their diesel engine vehicles.

Collaborative Biodiesel Tutorial
www.biodieselcommunity.org
How to make biodiesel.

Biodiesel
http://en.wikipedia.org/wiki/Biodiesel
From Wikipedia.

Biodiesel America
www.biodieselamerica.org
Book, news and information.

Biodiesel Basics
www.gobiodiesel.org/index.php?title=BiodieselBasics
Benefits and drawbacks.

Collaborative Biodiesel Tutorial
www.biodieselcommunity.org
Learn how to make biodiesel.

Diesel
http://en.wikipedia.org/wiki/Diesel
From Wikipedia.

Buy Carbon Offsets

See page 121.

TRANSPORTATION

Three Things to Do

1. Walk
2. Ride a bike
3. Use public transit

Details

1. WALK

It's free and healthy. See page 30.

2. RIDE A BIKE

Bicycles are one of the great inventions of the world. As John Ryan's book Seven Wonders: Everyday Things for a Healthier Planet says:

"The Bicycle: The most energy efficient form of travel ever invented and the world's most popular transport vehicle".

Pound for pound, a person on a bicycle expends less energy than any creature or machine covering the same distance. A bike is always handy for the one out of four car trips in the United States that are less than a mile.

Bicycle
http://en.wikipedia.org/wiki/Bicycle
From Wikipedia.

Bikeability Checklist
www.bicyclinginfo.org/cps/checklist.htm
How bikeable is your community?

Zap
www.zapworld.com
Zap sells electric bikes and kits for converting your standard bike into an electrified power assist bike.

Bike Web Site
www.bikewebsite.com
Bicycle tune-up and repair for all types of bikes, with lots of illustrations.

Jim Langley
www.jimlangley.net
Wonderful site on everything about bicycles, including buying one, using it and repairing it. From the former technical editor of Bicycling magazine.

> **Park Tool**
> www.parktool.com/repair
> Detailed instructions on bicycle repair and maintenance, with illustrations.

3. USE PUBLIC TRANSIT

If your community has public transit such as bus or light rail, use it. If it doesn't go where you need it to go, join with others to get it expanded or changed.

Want More?

Buy Carbon Offsets

Carbon offsets are environmentally beneficial actions that balance out the harm we cause by emitting carbon dioxide (CO_2) by travel, or through other causes. These beneficial actions are generally tree-planting or investing in renewable energy. They're basically "good karma" balancing out "bad karma". Look at carbon offsets as a kind of "environmental sin tax" or the equivalent of buying *indulgences* in the Middle Ages. You pay your penance and you get to keep doing the bad karma. (You might want to look at that, too, however, and determine how you can reduce your carbon emissions throughout your daily life—and *still* support tree-planting and renewable energy.)

Various websites let you buy "offsets" after calculating how much CO_2 you have to balance out to become "carbon neutral". For example, your "share" of an airline flight from San Francisco to New York is estimated to be as much as 1,000 pounds of CO_2.

> **Carbonfund.org**
> www.carbonfund.org

> Individuals and businesses can reduce their carbon footprint and support climate-friendly projects.
>
> **Native Energy**
> www.nativeenergy.com
>
> A privately-held Native American energy company. Their carbon calculator lets you calculate the carbon emissions resulting from your travel by auto, bus, rail or air. You can then buy "carbon offsets", offsetting those travel CO_2 emissions by financially supporting green energy projects that wouldn't happen otherwise.
>
> **TerraPass**
> www.terrapass.com
>
> Make your driving carbon-neutral by balancing your emissions with a TerraPass.

Electric/Gas Scooters

Scooters aren't just fun; they're inexpensive transportation. They get great mileage, using very little fossil fuel. A gallon of gas can can take you 50 miles or more. New scooters will cost you anywhere from $800 to $4,000, though it's possible to spend as much as $10,000.

> **Moped Army**
> www.mopedarmy.com
> Blog headquarters for moped clubs.
>
> **Motor Scooters**
> www.motorscooters.com
> Comprehensive information and sales.
>
> **Scooter**
> http://en.wikipedia.org/wiki/Scooter
> Wikipedia article.
>
> **Vespa**
> www.vespa.com

Official Vespa company site.

Electric Bikes
www.electric-bikes.com
Practical transportation for errands and short commutes.

Electric Vehicles

Full-size electric-only automobiles are not currently available, particularly since General Motors killed the EV1, but smaller electric forms of transportation are becoming more and more popular.

Battery Electric Vehicles
http://en.wikipedia.org/wiki/Battery_Electric_Vehicles
From Wikipedia.

EV World
www.evworld.com
Electric and hybrid vehicles.

Light Electric Vehicles
www.electric-bikes.com/lev.htm
From bikes and small scooters to one-person cars.

Segway
www.segway.com
The Segway Human Transporter.

ZapWorld
www.zapworld.com
Electric ATVs, scooters, and bikes.

Compressed Work Week

Compressed work schedules (such as a four-day, 40-hour work week) can eliminate commuting altogether one day a week for many employees. Companies with such programs report less absenteeism, fewer late employees, and less use of sick leave.

In a compressed work schedule program, employees work a full-time schedule in fewer days, by working more hours a day. The day off can be the same for all employees, vary or rotate regularly—but most employers choose to assign days off to ensure adequate coverage. The most common compressed schedules are:

"4/40" - A 40-hour week consisting of four 10-hour days and three days off a week

"9/80" - 80 hours worked over two weeks, consisting of eight 9-hour days, one 8-hour day and five days off.

Where can compressed work schedules best be used?

Compressed work schedules work best where employees require minimal face-to-face contact with other employees, where set-up/tear-down time or shift changeovers are necessary (e.g., hospitals or manufacturing), or where work functions are not disrupted by staff reduction.

Compressed Work Weeks
www.valleymetro.org/Rideshare3/9CWW
List of benefits.

BARTER

Three Things to Do

> 1. Barter with friends, neighbors and local businesses and services
> 2. Join, or start, a barter network
> 3. Develop a skill or product that you can use to expand your barter opportunities

Details

Barter is simply trading one thing for another. And, in fact, it doesn't even have to be a *thing*. It can be a skill, a service, or even information.

In a post-Peak Oil world, barter could be essential. Currency may have little value, while the value (usefulness, desirability) of certain things could increase dramatically. Now is a good time to begin to *demonetize*. Lessen your dependence on the financial system and its dollars, and increase your skills and relationships so that you can get the things and services you need by exchanging other things, services of your own, knowledge or anything else which might be considered of value. You'll learn that members of a community can help one another without needing to use pieces of paper (with pictures of dead presidents) to validate the mutual help.

Dealing with Peak Oil and economic collapse isn't about stocking up on stuff. It's about knowledge, skills, and cooperation. It's about being able to create and produce your own goods, either yourself or within your community. Nevertheless, some stuff is useful, and it doesn't hurt, no matter how good or bad the times, to have supplies of necessities around the house.

Remember that bartering doesn't have to be only in a one-on-one relationship. A mutual bartering system can be even more effective at the neighborhood or community level.

Running a barter network can also be an excellent form of self-employment. Barter is often confused with systems such as community currency. The difference is that barter involves only goods and services. Community currency, or systems such as LETS, involves either locally-printed currency or a computerized system to keep track of exchanges.

For closely related information, see Local Currency, page 138.

1. BARTER WITH FRIENDS, NEIGHBORS AND LOCAL BUSINESSES AND SERVICES

You should have no problem finding people you already know to barter with. It's really just an extension of giving away things you don't need. Local businesses might have more need of services rather than things. Ask and see what they can use.

2. JOIN, OR START, A BARTER NETWORK

Ask around to see if there is already a barter network in your community. There may actually be several. While there are statewide and nationwide barter exchanges, it's best if you can keep most of your bartering at the local level. It helps build community, and it cuts down on transportation expenses and use of fossil fuels.

3. DEVELOP A SKILL OR PRODUCT THAT YOU CAN USE TO EXPAND YOUR BARTER OPPORTUNITIES

The self-employment suggestions we provide can help you discover skills, products or services that you can use for barter. (See page 222).

Barter
http://en.wikipedia.org/wiki/Barter

From Wikipedia.

Small Lodge, the Great Depression, and Christmas
www.srmason-sj.org/council/journal/dec00/dodson.html

Small Masonic lodge in rural Virginia established a barter system to help its community through the hard times of the 1930s.

Barter - Relevance and Relation to Money
www.ex.ac.uk/%7ERDavies/arian/barter.html

Is barter still relevant in the modern world? Links and history.

Commerce Systems – Barter
http://collections.gc.ca/generalstore/commerce

A general store in frontier British Columbia. We may go back to those ways.

How to Barter
www.u-exchange.com/barter101.asp

Advice for online bartering but useful for face-to-face bartering as well.

Want More?

If you'd like to set aside high-demand items for later barter use, you'll find suggestions on page 202.

EXPENSES AND INCOME

Three Things to Do

1. Cut down on your expenses
2. Pay off debt
3. Find a self-employment opportunity

Details

Cutting down on your expenses saves money in many ways, and paying off debt goes even further since you're escaping from interest rates. Work continues in hard times, but you may need to be clever to find it. Ironically, at a time when so much more needs to be done, it gets harder and harder for people to actually find jobs.

We can guess what type of jobs might disappear first as a result of Peak Oil and economic collapse—such as housing construction, highway construction, just about any kind of construction for that matter, credit cards, banking, stocks and bonds, overnight tourism (daytripping may do okay), airlines (but bus and railroads should do well), the automobile industry and all its related businesses, hotels and motels, drive-through fast-food, car washes, chain stores dependent on cheap goods from China, and so on—but it's probably more useful to try and predict what type of jobs and skills will be *most* needed.

We assume they will involve basic needs—food, water, health, and the like. There will probably be much more focus on repairing items than replacing them, so cottage industries such as small appliance repair and clothing repair could do well. Escapist entertainment and fads thrived during the Great Depression of the 1930s and there's no reason to believe similar, low-cost businesses might not do equally well now.

For other related information, see Barter (page 125).

1. CUT DOWN ON YOUR EXPENSES

It's even better than increasing your income because you don't have to pay additional income and sales taxes.

Cut down on your use of electricity, gasoline and natural gas. This book is filled with suggestions on how to do that. Follow them, and you'll enjoy some major savings.

Grow at least some of your own food. For information on starting a home garden, see page 73. For a community garden, see page 62.

Buy less, sell more. Buy only what is absolutely essential, or what you *really*, really want. Look around your home and see if there are things you don't need—there is always someone, somewhere who is willing to buy those things.

2. PAY OFF DEBT

If you carried out action #1 above, you may be able to use some of the money you saved through cutting expenses to start paying off debt. Since much of any debt you have is liable to be credit cards and bank loans, the money you avoid having to pay in interest through lowering your debt is much more than you'd receive if you simply invested the money you saved from cutting expenses.

For example, if your credit card requires that you pay an 18% annual interest, paying off some or all of that card is the equivalent of investing the amount you paid at 18%. You'll have a hard time finding any investment that pays even close to that.

Best advice? Stop buying on credit unless it's an absolute emergency.

Here are some interesting figures about debt.

- In 2004, the credit card industry made $43 billion from fees for late payments, over-limit, and balance transfer.
- In 2005, total American consumer debt was $2.2 trillion, with an average of $11,840 per household.

- Average U.S. household credit card debt increased 167% between 1990 and 2004.
- In 2005, the average interest rate on credit cards was 14.5%.
- In 2005, the rate of personal savings in the United States was negative .5%, dropping below zero for the first time since the Great Depression.
- In 2004, 45% of U.S. cardholders were making only minimum payments on their credit card debt.
- As of 2004, a typical credit card purchase (including interest) is 12-18% more than if cash was used.
- In 2005, 2.39 million U.S. households filed for bankruptcy, a 12% increase over 2004.
- During the three years prior to 2005, 30 million Americans (40% of homeowners) refinanced their mortgages, with over half of them applying the proceeds to their credit card debt.
- In 2004, the average personal wealth of a 50 year-old American was less than $40,000—including home equity.

3. FIND A SELF-EMPLOYMENT OPPORTUNITY

Self-employment is no guarantee of financial security, but neither is the fact that you're currently employed. Look for work that can supplement or replace your current income (or lack of it). The one thing you can count on about the economy is that things are going to keep changing. Your job is to guess what those changes might lead to, and be there to meet them with new job skills, tools and whatever you need to take advantage of the opportunities.

- Do a self-inventory

Look at the skills, talents and experience you currently have and seek ways to turn those into an income.

- Study

Consider what skills you could learn in order to become self-employed, then take courses, read books and find other ways to learn them.

- Learn on-the-job

Get an entry-level job in the business you'd like to learn, or see if you can apprentice (even with no pay) with someone who already has these skills.

For specific suggestions for self-employment, see page 222.

Want More?

Consider How You'd Deal With Hyperinflation

If the financial situation in the United States continues as it has been going, the chances are great that the country will end up with very high inflation. This means that the purchasing power of your dollars—your paycheck, your investments, your retirement check—could become less and less; perhaps dramatically less. Now is the time to think about how you'd like to prepare for that eventuality.

The suggestions we've made above for getting out of debt are particularly important. The more you can pay down your mortgage, credit card and other debts—particularly those with variable interest rates—the less you'll be at the mercy of your creditors.

Invest In Yourself

If you're looking for somewhere to invest, why not start at home? Spend some money on things that will make you and your family as self-sufficient as possible. Buy seeds and equipment to start a garden. Buy bicycles. Buy compact fluorescent bulbs to replace your incandescent bulbs. Take classes that will increase your useful knowledge and skills.

Buy property for a larger garden, or for eventually relocation. Buy food storage equipment such as canning supplies and dehydrators. Buy solar panels or wind turbines. Buy equipment and supplies you'll need to set up your own business. Take classes in alternative health techniques. If you still have money, buy gold (see page 133).

Invest In Local Business

If you still have money you'd like to invest, consider assisting one or more local businesses. There are undoubtedly existing local businesses, or people wanting to start a business, that need a loan to start up or expand, or for other purposes. Some of these people may be unable to qualify for a standard bank loan, even though they have solid business plans. If you feel they're worth investing in, your loan, even if small, can make a significant difference.

You'll find it's possible to make loans locally at an interest rate higher than a bank will give you and, for the person borrowing the money, at a rate lower than they would get from a bank. It's a win-win for both of you.

For ideas on how this might work, take a look at Prosper.com, a website that brings borrowers and lenders together. They offer people-to-people lending, in which you can be the sole lender, or join with others so that each lender provides part of the loan needed by the borrower.

Through such investing, you're helping your community's economy as well as receiving a fair return on your investment. Making small investments like this is similar to, and based on, the programs initiated by the Grameen Bank, which began in 1976 in Bangladesh to give micro-loans to the poor who were unable to qualify for regular bank loans. The principle is similar here, although the monetary amounts are much greater.

Prosper
www.prosper.com
People-to-people lending.

Grameen
www.grameen-info.org
Micro-loans for the poor in Bangladesh.

GOLD AND SILVER

Three Things to Do

1. Buy gold
2. Buy silver
3. Don't sell them

Details

Historically, during hard times gold and silver have always become extremely valuable. Why should it be any different this time?

1. BUY GOLD

You can buy gold as coins or bullion (bars) from a local or online gold dealer, take delivery, and store it yourself. Or, you can buy from an online service that stores your gold securely for you.

E-Gold
www.e-gold.com

100% backed by gold. Also e-silver, e-platinum and e-palladium. Internet payments.

GoldMoney
www.goldmoney.com

Online transactions in gold and silver. Used for savings as well.

In Gold We Trust
www.wired.com/wired/archive/10.01/egold.html

From Wired magazine.

Kitco
www.kitco.com

Precious metals. Up-to-the-minute price charts.

Myths, Misunderstandings and Outright Lies
www.cmi-gold-silver.com/gold-confiscation-1933.html

Information about 1933 gold confiscation in the United
States and reportable gold sales.

Survival Coins
www.cmi-gold-silver.com/small-survival-gold-silver-
coins.html

Article on gold and silver bullion coins for survival
purposes.

What Most People Don't Know About Gold
www.gold-eagle.com/editorials_05/casey112305.html

From GoldEagle.com.

World Gold Council
www.gold.org

Organization formed and funded by the world's leading
gold mining companies to promote the demand for, and
holding of, gold by consumers, investors, industry, and
the official sector. Why, how and where to invest in
gold.

2. BUY SILVER

Silver is cheaper than gold. It's also possible to buy "junk"
silver from your local coin shop. These are U.S. coins that are
priced according to their silver content rather than by their
face value.

E-Gold
www.e-gold.com

> 100% backed by gold, e-silver, e-platinum and e-palladium. Internet payments.
>
> GoldMoney
> www.goldmoney.com
>
> Online transactions in gold and silver. Used for savings as well.

3. DON'T SELL THEM

Think carefully before you decide that you've "made enough" on your investment and sell your gold and silver for dollars. Although at times precious metals *can* be an investment, and can reward the speculative buyer, that's not what they're for.

Historically people have purchased gold—and to a lesser extent silver—as a way of storing value and, when needed, for use as exchange for goods and services. Gold retains its value relative to "stuff". For example, it is said that one ounce of gold would have bought you a good suit (probably a toga) in Rome, and that same ounce today would buy you a good man's suit in New York.

Where gold's value *does* change is in relation to currency, for example the U.S. dollar. But it's not gold that's changing in value, it's the dollar. Gold is an excellent investment in the future, as a means of storing its current value regardless of what happens to the dollar, the euro, or the yen.

LOCAL BUSINESS

Three Things to Do

> 1. Shop at locally owned businesses
> 2. Join with other locally owned businesses
> 3. Join (or start) a local currency system

Details

Local businesses help the local economy. Chain stores don't.

A study done in the Andersonville neighborhood of Chicago found:

- Local merchants generate substantially greater economic impact than chain retailers.
- Development of urban sites with directly competitive chain merchants will reduce the overall vigor of the local economy.
- Modest changes in consumer spending habits can generate substantial local economic impact.
- For every $100 in consumer spending with a local firm, $68 remains in the Chicago economy vs. $43 for spending at a chain store.
- For every square foot occupied by a local firm, local economic impact is $179 vs. $105 for a chain store

1. SHOP AT LOCALLY OWNED BUSINESSES

The people who own and work at those businesses are your friends, neighbors and fellow members of your community. They have the same stake in the community's success and health as do you. They're much more inclined to donate to local youth groups, schools and non-profit organizations than are chain stores. And the money they make is much more inclined to remain in the community, supporting other businesses, than is money that goes into the chain store vacuum to end up in a distant city.

If you own a business, it's particularly important that you use local and locally owned vendors, producers and manufacturers. Local businesses should support each other for the same reason that local customers should support local businesses. If you don't buy locally, why should you expect the residents of your community to buy from you?

2. JOIN WITH OTHER LOCALLY OWNED BUSINESSES

BALLE (the Business Alliance for Local Living Economies) and AMIBA (American Independent Business Alliance) are two national organizations that have local chapters supporting local business communities. You can help them spread their message of local economic independence, and they'll help you and your community's business interests. Check with them to see if there's already a chapter in your community. If not, start one.

Business Alliance for Local Living Economies (BALLE)
165-11th Street
San Francisco, CA 94103
415.255.1108
www.livingeconomies.org

Building long-term economic empowerment and prosperity in communities through local business ownership, economic justice, cultural diversity and a healthy natural environment.

American Independent Business Alliance (AMIBA)
222 South Black Ave.
Bozeman, MT 59715
406.582.1255
www.amiba.net

Locally owned independent businesses, citizens and community organizations united to support home town business.

Andersonville Study of Retail Economics
www.andersonvillestudy.com

The study quoted at the beginning of this section.

Benefits of Doing Business Locally
http://reclaimdemocracy.org/independent_business/local_
business_benefits.html

Benefits to communities and citizens in patronizing local businesses.

Buying Local and the Circulating Dollar
www.blueoregon.com/2005/11/buying_local_an.html

Often local business prices are lower than chain stores, and the money you spend there stays in the community.

Local Ownership Pays Off for Communities
www.reclaimdemocracy.org/independent_business/local_
ownership_pays.html

Financial benefits to the community of locally owned businesses.

The Home Town Advantage
www.newrules.org/retail

Reviving locally owned business - From the New Rules Project.

3. JOIN (OR START) A LOCAL CURRENCY SYSTEM

Local currency can have significant effects on the health of the local economy since it keeps money *within* the community (see below).

LOCAL CURRENCY

Three Things to Do

1. If there is no local currency system, start one
2. Whenever there's a choice, patronize a business that accepts local currency
3. If you have a business or skill, make it available to the public using local currency

Details

Money is the central focus of our society and it's extremely difficult to get away from that focus. Some try to work with money in ways that benefit a community rather than just individuals. Networks supporting locally owned businesses (page 137) are one of these ways; "community currencies" are another.

Many people have no faith in the government's currency since it can be printed by the government in quantities with no apparent limitations, and has no real backing in reality. It is "faith-based".

Local Currency, also called Community Currency, is paper currency printed and distributed locally, of value only within a community. This currency remains in the community instead of being removed by distant owners. It circulates from person to person, business to business, benefiting the entire community.

Local currency can also empower citizens who may not normally be involved in the economic life of their community. And it's popular when no one trusts the national currency, as was seen throughout the United States and other countries during the Great Depression of the 1930s.

If you were in the U.S. military in Vietnam, you used local

currency. There it was called "Military Payment Certificate (MPC)", usable only at U.S. installations. If you've ever been to a Disney theme park, you probably used "Disney Dollars". That's also local currency, although it benefits a corporation and not a community. (A rumored 60% of the currency is kept as souvenirs and never spent, resulting in nice profits to Disney just for printing paper.)

Local currency, which involves use of locally printed money (yes, it's legal), is often confused with *barter*, which involves only goods and services and no currency. We recommend you also look into barter (see page 125), which quite nicely complements the use of local currency.

1. IF THERE IS NO LOCAL CURRENCY SYSTEM, START ONE

Talk to small, locally owned businesses, particularly health food stores, bookstores, farmers markets, and cooperative stores. Many of these businesses already appreciate the value of community and cooperation. A locally owned bank may also understand the value of local currency.

Gather a small group together and start small, expanding as you feel more comfortable with the system.

Complementary Currency Resource Center
www.complementarycurrency.org

Information, resources, step-by-step instructions, online assistance, and forums.

Starter Local Currency Kit for your community
www.lightlink.com/hours/ithacahours/starterkit.html

Hometown money book and samples of money.

---

2. WHENEVER THERE'S A CHOICE, PATRONIZE A BUSINESS THAT ACCEPTS LOCAL CURRENCY

Patronizing businesses that accept local currency supports those businesses and encourages them to continue their policy. Let other businesses know you'll patronize them also, once they start accepting the currency.

3. IF YOU HAVE A BUSINESS OR SKILL, MAKE IT AVAILABLE TO THE PUBLIC USING LOCAL CURRENCY

Local currency provides an excellent way for individuals to integrate into the local economy. Your individual services can be listed in the local currency members directory right along with established businesses in town.

Community Currencies
www.transaction.net/money/cc/cc01.html

How local currency can help solve the problems of unemployment, the environment, and community breakdown. The philosophy, history and potential for the 21st century.

Complementary Currencies
www.transaction.net/money/community/

Extensive information and resources on complementary community currency systems and local exchange networks.

E.F. Schumacher Society Home Page
www.schumachersociety.org

Linking people, land, and community by building local economies. Named after the author of *Small Is Beautiful: Economics As If People Mattered.*

E.F.Schumacher Society's Local Currency Page
www.smallisbeautiful.org/local_currencies.html

Fair Trade
www.newdream.org/newsletter/fairtrade.php

On community currency and other systems.

Grassroots Economics
www.context.org/ICLIB/IC41/Glover.htm

Local currency in Ithaca, New York.

Ithaca HOURS
www.ithacahours.org

Detailed information on the Ithaca community currency.

Ithaca HOURS Online
www.lightlink.com/hours/ithacahours

Locally issued paper money from a well established community currency project in Ithaca, New York.

LETSystems
www.gmlets.u-net.com

Lets (Local Employment Trading System) is a trading network supported by its own internal currency. It is self-regulating and allows its users to manage and issue their own "money supply" within the boundaries of the network.

Barter/LETS 101
www.cyberclass.net/bartable.htm

Usury-free software information for LETS and LETS communities around the world.

Let's Go Global with Barter
www.alternatives.ca/article138.html

But actually about LETS.

More on LETS
www.transaction.net/money/lets/

From Transaction.net ncluding links to communities using the system.

Self-sufficiency Economy by Own Money
www.appropriateeconomics.org/asia/thailand/
self_sufficiency_economy.htm

Community barter [currency] system in a Thai village.

Time Dollar
www.timebanks.org
A time-based currency where one hour spent helping
another earns one Time Dollar.

BATHROOM

Three Things to Do

1. Turn the faucet off
2. Use a low-flow toilet
3. Use a low-flow shower head

Details

The best ways to conserve in the bathroom are by using less water and particularly less hot water (see page 146).

1. TURN THE FAUCET OFF

When brushing your teeth or washing your face, turn the water off except when you actually need it. You can save several gallons of water each time.

2. USE A LOW-FLOW TOILET

If you have a regular toilet, replace it with a low-flow toilet. Many municipalities offer free replacements, including the labor. At the very least, put in water displacement bags to lessen the amount of water the tank actually holds.

3. USE A LOW-FLOW SHOWER HEAD

Replace any normal shower heads with low-flow. They're inexpensive, and some municipalities even give them away for free. You can reduce water usage by almost 50% and save water-heating money as well.

Want More?

Don't Flush

For urine only, it's not necessary to flush each time. Save the water for when it's essential. As they used to say in California during the 1970s drought, "Brown goes down, but yellow is mellow."

Shower Together

They liked this one in California.

HOUSEHOLD

Three Things to Do

> 1. Replace incandescent light bulbs with compact fluorescent bulbs
> 2. Disconnect all electrical items when not in use
> 3. Turn down thermostats and water heater temperature

Details

After the mortgage or rent, gas and electricity are likely the biggest expenses in your home. Do the following three things and you'll have taken the biggest and easiest cost-cutting measures that you can take in your home.

1. REPLACE INCANDESCENT LIGHT BULBS WITH COMPACT FLUORESCENT BULBS

See page 29.

2. DISCONNECT ALL ELECTRICAL ITEMS WHEN NOT IN USE

Few people realize just how much electricity many appliances can consume when they're not actually being used but are still plugged in. Such appliances are called "energy vampires" or "phantom appliances". These include radios and television sets, VCRs, home entertainment systems, battery chargers (for cell phones, cameras, music players), computers, printers, photocopiers, and even kitchen appliances.

The U.S. Department of Energy estimates that 75% of the electricity used to power home electronics is consumed *while the products are turned off*. This can be avoided by unplugging the appliance, or by using a power strip so that when you turn the power strip off you cut all power to the appliance.

Obviously you'll want to keep some appliances plugged in if, for example, they're dependent on maintaining an accurate clock to carry out their function and they don't have an internal battery to do that. But others can be unplugged between use.

To find out how much electricity each of your appliances consumes, even when it's not actively being used, we suggest that you buy a *Kill-a-Meter*, an electricity usage monitor.

3. TURN DOWN THERMOSTATS AND WATER HEATER TEMPERATURE

Thermostat

Reducing temperatures saves electricity and/or gas and the fossil fuels required to produce them. It also saves you money. On the average, every 2° F. lower on your thermostat will save 2% on your heating bill. Conserve further by turning down the heat at night and while you are away from your home—or install a programmable thermostat.

Water Heater

The water heater uses more energy than any other device in the home. Most water heaters are kept at the factory-set temperature of 140°-145°. The expense of maintaining a tankful of water at this temperature, 24 hours a day, is very high. The high temperature also a safety hazard if you have small children in the home.

Turn the temperature down to 120°, your water will still be hot enough for showers, dish washing (most new automatic dishwashers raise the temperature themselves, if necessary), and clothes washing. The lower temperature will use a lot less gas or electricity, and save you a lot of money. If, for some reason, you find that 120° is too low for your purposes, you can always set it a little higher.

Note: An even better solution, but one which requires an expense, is converting your always-on water heater to a tankless device that heats water on-demand. These have been standard in Europe and other parts of the world for decades, and are slowly becoming common in the United States.

Want More?

- **In The Winter, Change Or Clean Your Furnace Air Filters Once A Month.**

The heater has to use more energy when the filter is full of dust.

- **Insulate Your Home Against Heat Loss And Periodically Check The Insulation.**

- **Weather-strip And Caulk To Block Any Air Leaks.**

- **Turn Your Computer Off When You're Not Using It.**

It will reduce wear on the computer, and save energy and money. If you have to leave the computer on, at least turn the monitor off. CRT monitors use more than half of the system's energy (LCD monitors are much better).

Laptop computers use up to 90% less energy than a desktop computer.

Home Energy Saver
http://hes.lbl.gov/
Do-it-yourself energy audit tool from Lawrence Berkeley National Laboratory.

KITCHEN

Three Things to Do

1. Unplug appliances not being used
2. Switch to hand-operated appliances and tools
3. Use solar cooking whenever possible

Details

1. UNPLUG APPLIANCES NOT BEING USED

See page 145.

2. SWITCH TO HAND-OPERATED APPLIANCES AND TOOLS

Get rid of the electric juicer, can opener, and whatever else you have that does what could be done almost as easily by hand.

3. USE SOLAR COOKING WHENEVER POSSIBLE

See page 67.

Want More?

- Don't preheat your oven. That's necessary only for baking breads and pastries.
- Turn off the stove burners 2-3 minutes (and the oven 10 minutes) before the end of cooking time. The food will continue to cook and you'll save money.
- Electric kettles use much less energy than a stove burner when heating water. Clean the kettle regularly with boiling water and vinegar and you'll remove the mineral deposits inside, improving the taste of the water and making the kettle more energy efficient.
- Use small appliances such as toaster ovens or microwaves to cook or re-heat small amounts of food. They'll use up to 50% less energy than your oven.
- Skip the pre-rinse for your dishwasher and you'll save up to 20 gallons of water per load.
- Scrape off dishes instead of rinsing them. Most new dishwashers can handle this. Run the dishwasher only when it's full.
- Your refrigerator uses more energy than any other home appliance. Keep your refrigerator coils clean (at least every year and preferably every six months), don't keep the refrigerator jam packed with food (although this helps the freezer), and keep the refrigerator set between 37°-42° F. and the freezer between 0° and 5° F.
- Water boils faster if there's a lid on the pan. Once it's boiling, turn it down to a light boil instead of a rolling boil. The temperature will be just as hot.
- Thaw frozen foods before cooking; they'll need less energy to cook.
- Food cooks faster in glass dishes than metal ones.
- Pressure cookers use very little energy and cook very quickly. If you're afraid of them from earlier days,

don't worry. The new ones are much easier—and safer —to use.

LAUNDRY ROOM

Three Things to Do

> 1. Set water heater to 120° F
> 2. Wash laundry with full loads
> 3. Use a clothesline

Details

1. SET WATER HEATER TO 120° F

See page 146.

2. WASH LAUNDRY WITH FULL LOADS

The optimal and most economical use of water and electricity is to wash only full loads of laundry ("full" as determined by the washer manufacturer; don't cram the laundry in the washer). Smaller loads use just as much electricity, and often more water than necessary, even with water volume set on a partial amount.

3. USE A CLOTHESLINE

Use a clothesline rather than an electric or gas dryer. If the weather outside isn't warm enough to dry your clothes, use an indoor clothesline or clothes dryer rack. You'll save money on energy, and your clothes will last longer.

Want More?

- Wash clothes in cold water. You'll save on water heating costs, and studies show clothes get just as clean in most washers.
- If you use a clothes dryer, dry loads of clothing consecutively. This way you take advantage of the heat from the previous load. Be sure to clean the lint trap after each use.

SANITATION

Three Things to Do

1. Switch to compost toilets
2. Reuse your graywater
3. Use alternatives to toilet paper

Details

Given any thought to what you'll do if your local sewage system doesn't work? Never fear. There are ways—without wasting what's valuable.

1. SWITCH TO COMPOSTING TOILETS

With one flush of your toilet you're using more water than a majority of people in the world have access to in an entire day. And in most cases, you're flushing not just water, but *drinking* water.

Composting toilets save huge amounts of water, since they never use *any*. The human manure ends up as compost—in a safe, sanitary and non-smelly method.

Composting Toilet
http://en.wikipedia.org/wiki/Composting_toilet

From Wikipedia.

Composting Toilet World
www.compostingtoilet.org

Extensive information from Envirolet, a manufacturer of composting toilets.

Build Your Own Humanure Toilet
www.jenkinspublishing.com/sawdustoilet.html

Complete instructions for a $25 toilet.

Humanure Handbook
www.weblife.org/humanure

A guide to composting human manure. The entire book online—free! Great book, great deal. Consider, however, buying a copy of the book for convenience and to support the author.

The Toilet Papers

Book on recycling waste and conserving water. A classic that's back in print. It's by Sim van der Ryn, well-known sustainable architect and former official California State Architect.

2. REUSE YOUR GRAYWATER

Graywater is the water from dishwashing, laundry, showering and bathing. While no human wants to drink it, plants are happy to get it. You can set up a system that will save and recycle that water to be used in your garden.

Graywater Central
www.graywater.net
All about graywater.

3. USE ALTERNATIVES TO TOILET PAPER

The following aren't so much recommendations as they are suggestions in case you have to deal with a *lack* of toilet paper.

Hint: The Romans used a sponge attached to the end of a stick which soaked in a bucket of brine, early Americans used corncobs and, later, newspapers. Caution: Corncobs and newspaper would not be handled well by a flush toilet.

Toilet Paper
http://en.wikipedia.org/wiki/Toilet_paper
An overview of toilet paper, from Wikipedia.

The Hand/Water Method
www.getlostmagazine.com/features/2000/0004hygiene/hygiene.html
The standard method used throughout much of the world.

Investigating TP Alternatives
www.clarkschpiell.com/home/asswipe.shtml
Some imaginative, and even amusing, alternatives.

Where's the Toilet Paper?
http://bosp.kcc.hawaii.edu/DiamondJournals/Diamondspring03/Wheres_TP.html
A commentary.

The Great Toilet Paper Shortage
http://home.nycap.rr.com/useless/toilet_paper/index.html
Interesting historical information.

SHELTER

Three Things to Do

> 1. Share housing
> 2. Green your home
> 3. Build green

Details

In hard economic times, shared housing, in one form or another, offers perhaps the most sensible and easiest way to cut costs.

Greening your home, that is, making it healthier and more sustainable, saves money and energy.

Building homes of natural, sustainable materials, designed and engineered for minimal energy use and maximum comfort, makes both economic and aesthetic sense.

1. SHARE HOUSING WITH OTHERS

Shared housing decreases the cost (and use of resources) per person, and increases the opportunities for social interaction and shared work and responsibility.

There are a number of types of "sharing" housing. Even the nuclear family home can be seen as shared housing—although an "extended" nuclear family with as many people as you can handle makes more sense. Boarding houses (also called "rooming houses") are another way people share housing.

Communal living, where a group of perhaps unrelated people live together in one house, or in a cluster of houses has been popular for a long time.

There are also ecovillages and intentional communities. You can find more information on them at page 50.

One of the most interesting forms of shared housing is

"cohousing", where people don't share homes but do share common facilities.

Cohousing

Cohousing is collaborative housing that attempts to overcome the alienation of modern subdivisions where people don't know their neighbors, and there is no sense of community.

It's characterized by private dwellings with their own kitchen, living-dining room etc, but also extensive common facilities. The common building may include a large dining room, kitchen, lounges, meeting rooms, recreation facilities, library, workshops, childcare.

Usually, cohousing communities are designed and managed by the residents, and are intentional *neighborhoods*: the people are consciously committed to living as a community; the physical design itself encourages that and facilitates social contact.

The typical cohousing community has 20 to 30 single family homes along a pedestrian street or clustered around one or more courtyards. Residents of cohousing communities often have at least several optional group meals in the common building each week.

This type of housing began in Denmark in the late 1960s, and spread to North America in the late 1980s. There are now more than 80 cohousing communities across the continent, with many more in progress.

Cohousing [book]
Author: Kathryn McCamant
A contemporary approach to housing ourselves.

Senior Cohousing [book]
Author: Charles Durrett
Designing cohousing communities for senior living.

Cohousing Association of the United States
www.cohousing.org

Community list, products and services, resources, news.

Cohousing Company
www.cohousingco.com
The people that pioneered cohousing in North America.

2. GREEN YOUR HOME

Peak Oil Prep is filled with ideas on how to make *living* in your home more sustainable. But there are also many things you can do to make your home *physically* more sustainable. With a number of physical changes, renovations and additions to your home, it can be healthier, more cost-efficient, less demanding of resources, and a more pleasant place to live.

We don't have room to list the many things you can do, but we do have room to highly recommend this book:

Natural Remodeling for the Not-So-Green House [book]
Author: Carol Venolia

Greening begins at home. How to renovate your home to make it more environmentally-friendly. From simple to large-scale remedies.

3. BUILD GREEN

It's best if we stop building homes. Although we need homes, we have enough buildings. We need to *renovate* the buildings we have and turn them into healthy, comfortable, energy-efficient homes.

That said, if you're going to go ahead and build a home, make sure that at least you've done everything you can in its design, siting, and materials to make it as healthy and sustainable as possible. (And if you're a developer putting up a whole bunch of homes, please *don't* do *cul de sacs*, and please *do* include locally-owned stores, restaurants and other services in each neighborhood. If the city won't let you do mixed-use development, don't do the development.)

A Pattern Language: Towns, Buildings, Construction [book]
Author: Christopher Alexander

Architect Christopher Alexander's almost legendary opus on the elements that make a building or an entire town truly human.

Pattern Language
www.patternlanguage.com

We also highly recommend Alexander's website. You'll find lots of information and resources, including his most recent books.

Alternative Construction [book]
Author: Lynne Elizabeth

Contemporary natural building methods.

The House to Ourselves [book]
Author: Todd Lawson

Reinventing home once the kids are grown. Excellent resource for baby boomers and other empty-nesters, including those who want to live with friends in homes especially designed for groups..

Living Homes [book]
Author: Thomas J. Elpel

Sustainable architecture and design.

The New Ecological Home [book]
Author: Dan Chiras
A complete guide to green building options.

The New Natural House Book [book]
Author: David Pearson

Creating a healthy, harmonious and ecologically sound home.

Alternative Building Materials

Earthship

Tires + dirt = housing. Now doesn't that seem like a good way to recycle the millions of junked tires scattered around the country?

> **Earthship**
> www.earthship.org
>
> Recycled automobile tires filled with compacted earth to form a rammed earth brick encased in steel belted rubber.

Rammed Earth

They're just dirt houses. How can they hold up? Actually, homes made of rammed earth hold up fine, and have for centuries. New techniques of construction make them even more solid and easier to work with, and they provide year-round insulation, quiet and comfort. Plus, they just *feel* good.

> **Rammed Earth Works**
> www.rammedearthworks.com
>
> Pioneering California company that also invented the PISÉ (Pneumatically Impacted Stabilized Earth) process.
>
> **The Rammed Earth House [book]**
> **Author: David Easton**
>
> Excellent book by rammed earth pioneer.

Straw Bale

"But can't wolves blow down houses made of straw? I seem to remember a story about that." Not these. They're solid. And easy to shape into the type of structure you want. A great way to recycle straw.

> **The New Straw Bale Home [book]**
> **Author: Catherine Wanek**

Good coffee table book that will leave you wanting to build your own.

Serious Straw Bale [book]
Author: Paul Lacinski

A Home Construction guide for all climates.

Strawbale Central
www.strawbalecentral.com

Information on many different natural building techniques.

Strawbale.com
www.strawbale.com

Good information and resources including DVDs.

YARD

Three Things to Do

1. Plant a garden
2. Plant trees
3. Convert your lawns into gardens

Details

1. PLANT A GARDEN

See page 73.

2. PLANT TREES

Particularly nut and fruit trees (See page 103).

3. CONVERT YOUR LAWNS INTO GARDENS

The best thing you can do is rip out your lawns—which are heavy users of water—and replace them with gardens. Food is more important than trying to pass off your home as an English manor house.

There are three basic methods for removing a lawn:

- ◆ Use a hoe and spade to scrape away the turf, then turn the soil.
- ◆ Use a spade and remove the turf in pieces, cutting roots along the way.
- ◆ Cover the grass with newspaper, then cover the paper with six inches or so of topsoil. Some months later the grass will have died off and decomposed. (This method is obviously easier, but takes much longer.)

> **Converting Lawns to Gardens**
> www.backyardnature.net/simple/lawn2gar.htm
>
> **Lawns to Gardens**
> www.yougrowgirl.com/lawns_gardens_convert.php

Want More?

Use A Hand Lawn Mower

If you're not converting all your lawns into vegetable gardens, at least switch over to a hand-powered lawn mower. You'll save money, get some exercise, and reduce energy use and air pollution. If you can't use a hand mower, at least switch over to an electric mower. They don't pollute, and are a lot quieter than power mowers.

Gas-powered lawn mowers are a major source of air pollution. So much so that many municipalities have programs to help you replace your gas mower with an electric one.

COOLING

Three Things to Do

1. Use window coverings
2. Use ceiling fans instead of air conditioning
3. Shade your home

Details

1. USE WINDOW COVERINGS

Close drapes, curtains, blinds and windows on south and west-facing windows on hot days. This is particularly important with south-facing windows but should be done with all windows. At night, open them up to let the cool air in (if there is any).

2. USE CEILING FANS INSTEAD OF AIR CONDITIONING

Although they don't actually lower the temperature of the air in the room, they lower the *perceived* temperature. Fans move the air around creating a wind chill effect that makes you *feel* cooler because of increased evaporation of moisture on your skin. You'll save a lot of energy by using fans instead of air conditioning.

Remember that since they cool *you* and not the air, there's no point in having them on when no one's in the room.

3. SHADE YOUR HOME

Use trees, deciduous plants and/or awnings to shade your home and windows. For more, see page 103.

Want More?

♦   If you must use air conditioning, keep your thermostat set no lower than 77° F. when you're at home, 85° F. when you're away from home. Your air conditioner uses three to five percent more energy for each degree below 75° F. Set it to 77° F. for the most comfort at the least cost.

♦   Turn off your furnace pilot light during the warm season. You'll save both money and energy. For safety, your utility company will probably do this at your request.

LIGHTING

Three Things to Do

1.   Use compact fluorescent bulbs
2.   Turn off lights when you don't need them
3.   Get up earlier and go to bed earlier

Details

1. USE COMPACT FLUORESCENT BULBS

See page 29.

2. TURN OFF LIGHTS WHEN YOU DON'T NEED THEM

Leaving lights on in an unoccupied room is a waste of electricity and money. Turn the lights off when you leave, and back on when you return. Despite what many people believe, the act of turning a light on uses very little electricity. It really does save electricity and money to turn them off.

3. GET UP EARLIER AND GO TO BED EARLIER

Get up at sunrise and go to bed earlier in order to take advantage of natural light. Natural light saves energy and is easier on your eyes, you'll be more in sync with natural rhythms, and you'll save money on lighting.

HEATING

Three Things to Do

> 1. Set your thermostat lower than normal
> 2. Replace/clean your furnace filter regularly
> 3. On sunny days, use passive solar heating

Details

More than half of the energy used in the average home is on heating and cooling. Remember that your goal doesn't have to be to heat the entire home; it's simply to heat the people in it.

Dressing warmly can be a major way of dealing with the cold; putting on a sweater is cheaper than raising the thermostat. Devices such as radiant heaters are also good because they use much less energy since they heat people, not the air in the room.

1. SET YOUR THERMOSTAT LOWER THAN NORMAL

Keep your thermostat set no higher than 68° F. during the day. Wear warm clothing if necessary. Set the thermostat to 55° F. before going to bed at night.

2. REPLACE/CLEAN YOUR FURNACE FILTER REGULARLY

A clean-flowing filter will be more effective and require less energy than a dirty filter.

3. ON SUNNY DAYS, USE PASSIVE SOLAR HEATING

Keep the drapes open on south-facing windows to let the sun

shine in. At night, close the drapes to retain indoor heat.

Want More?

Ceiling Fan

During the winter, set your ceiling fan to run clockwise. Since hot air rises, the fan will help push the warmer air to the ceiling and then down the walls, circulating the warmer air at lower levels.

Stay Warm At Night

We lose most of our heat at night through our head and neck. In addition to warm blankets and warm pajamas or other night clothes, wear a nightcap. Socks or other comfortable foot coverings will keep your feet warm as well.

POWER

Three Things to Do

1. Use less
2. Generate your own
3. Use a solar battery charger

Details

Power is what makes just about everything work. Producing it on a large scale requires lots of fuel, such as coal, oil and natural gas. If there's a shortage of any of those fuels, serious problems can, and will, begin.

But you can produce your own energy, in ways that are sustainable, economic (at least in the long run), and most importantly, available.

1. USE LESS

The three most important areas in your home where you can cut electricity use and save money are:

LIGHTING

Replace incandescent bulbs with compact fluorescent bulbs (see page 29).
Turn off lights when you're not using them .

HEATING

Turn down your heating thermostat (see page 145).
Turn down your water heater (see page 145).

COOLING

Turn up your air conditioner thermostat (see page 160).

Saving Electricity
www.michaelbluejay.com/electricity

An excellent website with very specific information on how to save money on your home appliances and systems.

Kill-a-Watt

An "electricity usage monitor" that measures the amount of electricity consumed by household appliances—even when they're not actually being used but are simply plugged in. A great way to see where you can reduce your electricity consumption. [product]

---

2. GENERATE YOUR OWN

Depending on your climate and location, there are a number of options for generating your own power. If you're fortunate, you might be able to combine two or three of these methods to ensure year-round power.

Solar Power

Solar is the most common form of renewable energy. The sun doesn't have to be shining brightly to produce energy, but it helps.

Solar requires a high initial expense because of the cost of photovoltaic cells, which convert sunlight into electricity. If you're fortunate, you live in an area where your state or utility gives rebates, or at least loans, to install solar.

Real Goods
www.realgoods.com

The pioneer in home solar energy.

Real Goods Solar Living Sourcebook [book]
Author: Real Goods

The classic information catalog.

Small Solar Electric Systems
www.eere.energy.gov/consumer/your_home/electricity/
index.cfm/mytopic=10710

From the U.S. Department of Energy.

Wind Energy

Wind is better than solar if your area has enough reliable wind. Wind turbines of various designs turn from the force of the wind and generate electricity. A number of firms have the goal of producing low-cost wind turbines for the home of the size of a satellite dish. Unfortunately they're not perfected yet.

Small Wind Electric Systems
http://www.eere.energy.gov/consumer/your_home/electrici
ty/index.cfm/mytopic=10880

From the U.S. Department of Energy.

Wind Energy Basics [book]
Author: Real Goods

A guide to small and micro wind systems.

Wind Power, Revised Edition [book]
Author: Paul Gipe

Renewable energy for home, farm, and business.

Hydro Energy

Hydro, or water, energy is best known from the large dams that block rivers and generate huge amounts of energy from the water's flow—or fall. Micro-hydro, which produces up to 100kW, is the mini-version of hydro for the home. It's the most efficient form of sustainable energy generation and is ideal if your home is right next to a year-round stream.

Pico-hydro generates up to 5kW of electricity. Low-cost (less than $US 20) water turbines from China are currently very popular in rural areas of Vietnam.

Micro-Hydro Power Systems
www.eere.energy.gov/consumer/your_home/electricity/index.cfm/mytopic=11050

From the U.S. Department of Energy.

Pico Hydro
www.picohydro.org.uk

Network promoting small hydro systems up to 5kW.

All About Hydraulic Ram Pumps [book]
Author: Don Wilson

The device that can pump water from a flowing source of water to a point higher than that source using the force of gravity rather than power.

How to Live Without Electricity - and Like it [book]
Author: Anita Evangelista

> **Live off the grid but still have power, water, heating, and refrigerated food.**
>
> **Power with Nature [book]**
> **Author: Rex A. Ewing**
>
> **Highly-recommended book on solar, wind, micro-hydro, heating house and water and pumping water.**

3. USE A SOLAR BATTERY CHARGER

With a solar battery charger and chargeable batteries, all you need is sunshine to keep all your battery-operated devices working. They're available online and at a variety of local shops such as hardware and camping stores.

Want More?

Buy A Wind-up Or Solar-powered Radio.

There are a number of radios on the market that are charged up by turning a hand-crank, or by using a small solar panel to charge the battery. Some models include both methods. You'll never again need batteries or an electrical outlet to listen to the radio.

> **FreePlay**
> **www.freeplayenergy.com**
>
> **The pioneer in wind-up, solar and rechargeable technology.**
>
> **C. Crane**
> **www.ccrane.com**
>
> **Excellent source of radios and other electronics.**

REDUCE / REUSE / RECYCLE

Three Things to Do

> 1. Reduce
> 2. Reuse
> 3. Recycle

Details

Use the Three Rs—Reduce, Reuse, Recycle—when dealing with *stuff*.

1. REDUCE

Reduce the amount of stuff that you buy. Simply use less. Buy only what you need—or what you *really*, really want. Buy tools, appliances, furniture, clothing and other items that are designed to last for many years. Avoid buying things designed to be quickly used and thrown away. Seek out items that were made with sustainable materials under appropriate labor conditions.

"Reduce" is also referred to as "precycling". Avoid stuff in the first place. Wherever possible, avoid packaging and containers. If you have to get them, get containers that are recyclable. Avoid prepackaged and prepared foods. Buy whatever you can in bulk so save packaging and reduce costs, and prepare your meals from scratch.

Precycling means avoiding waste (and often saving money) by making good environmental decisions at the store.

Select products in recyclable containers such as paper, cardboard, glass and aluminum. Avoid disposable and single-use products unless absolutely essential.

Avoid plastic containers whenever possible. If you have no choice, make sure that the recycling symbol (three arrows in a circle) is on the packaging.

2. REUSE

Avoid disposable items. Buy items that can be used over and over; if not by you, then by someone else. Buy items that can be maintained and repaired so that their useful lifetime can be extended.

3. RECYCLE

When you finally have no need of something, pass it on to someone else who can use it. If it is past the point where it can be used, recycle it so that its materials can be reused in some other form.

Recycle everything made from metals, plastic, wood, fibers, glass or paper. Someone, somewhere, probably has a use for it. Clothing, furniture, household items, tools and toys can all be used by someone else, particularly if they can get them for free.

Books can go to the local library for use or resale. Usable items can be given to local charities and thrift shops, taken to local materials recycling depots, or simply put in front of your house with a big sign that says "Free" (if you local regulations allow).

SHOPPING

Three Things to Do

1. Avoid prepackaged products
2. Carry your own cloth shopping bags
3. Skip the middle of the supermarket

Details

1. Avoid prepackaged products

A significant percentage of the cost of many products is the packaging. The more you can avoid packaged products—particularly frozen foods, snacks, prepared mixes, and the like—the more money you save and the more you avoid the need for recycling (see page 168).

2. Carry your own cloth shopping bags

Carry your own sturdy cloth bags for shopping. You will no longer have to make the "Paper or plastic?" choice (which, in essence, is "Tree or oil?"). It avoids any need for recycling, and cloth bags are a lot stronger than paper or plastic bags. Most grocery stores sell them; some even give them away.

3. Skip the middle of the supermarket

In most supermarkets, that's where all the prepared foods and snacks with additives, preservatives and heavy sugar are. On the perimeter are the real foods: meat, dairy and produce. Venture into the middle only for such essentials as flour, cooking oil, spices, cleansers and the like.

Want More?

Avoid All Chain Stores

Shop locally (see page 136). If you still have them, patronize local specialty stores such as butchers, fishmongers, green grocers, hardware stores, office supplies and the like.

FAMILY

Three Things to Do

> 1. Eat together
> 2. Entertain together
> 3. Hold weekly meetings

Details

1. EAT TOGETHER

One of the saddest things about current American life is that less than half of families have—or take—the time to eat together. Whether because of work, sports, friends, entertainment, shopping or second jobs, many people eat on the run, often alone or elsewhere, such as at fast-food restaurants.

Eating together as a family, usually at dinner, is not just some nice old-fashioned idea that might have worked fifty years ago but is no longer appropriate for today. The family— and by this we mean those living together in a home, whether blood-related or not—is a core unit of society. The more stable the family is, the more stable society is. It's clear that both are pretty shaky these days.

Cook together, eat together, talk together, listen to one another, support one another. It will make a difference. And help make your home a special place for all of you.

2. ENTERTAIN TOGETHER

In the 1930s in the United States, families would gather around the *one* radio in the home and listen to comedy, news and drama programs. Today you may have a TV in every room, but that doesn't mean there isn't value in coming together and doing something together at least once a week.

We don't suggest that the family comes together by *edict*, but rather by *attraction*. Find things that family members actually enjoy and want to do. We'll leave the possibilities up to you, but they might include such things as playing cards and games, listening to—or making—music, reading books aloud or discussing current events. You can probably come up with more creative ideas based on the interests of your own family.

Does this sound corny to us jaded 21st century people? Maybe. Is it still a good idea? You bet.

3. HOLD WEEKLY MEETINGS

We're not talking about a meeting where Mom and Dad gather everyone together, pretend to take input from the kids, assign jobs, and then dismiss everyone after having a cup of hot cocoa. (Actually, the cocoa's not a bad idea.)

The reality is that a huge number of homes don't have both Mom and Dad, and many others don't have either. But it's still important that everyone that *does* live in the household gathers together weekly to discuss how the household is going, to clear the air of any gripes and misunderstandings, and to set goals that will benefit everyone. Let family members take turns running the meetings to lessen the possibility of any one person always controlling them, and to ensure that there is a variety of opinions and creative ideas. It will also give you kids good experience in running meetings.

KIDS

Three Things to Do

1. Turn off the TV
2. Get them to read
3. Teach them a skill

Details

1. TURN OFF THE TV

Save the electricity and get them outside where they can get some exercise in the fresh air. If you can't get them to quit cold turkey (quite likely), at least drastically cut down on the hours viewed each week.

2. GET THEM TO READ

Other than for lighting at night, reading requires no use of electricity. It entertains, stimulates the mind and improves one's ability to see possibilities in the world. Read to them, read with them, let them read to you.

3. TEACH THEM A SKILL

Any skill. Learning one thing will give them the confidence to learn another. And another. Focus on skills they enjoy, but try to veer them toward skills that would also be useful in a world of less oil and energy.

PETS

Three Things to Do

1. Grow your own catnip
2. Make your own pet food
3. Heal your pets holistically

Details

1. GROW YOUR OWN CATNIP

Fluffy needs a little stress relief and entertainment, too. You can grow all she needs.

Grow Your Own Catnip
http://home.ivillage.com/pets/cats/0,,hkp2,00.html

2. MAKE YOUR OWN PET FOOD

Save money and give them healthier food. For example, here's a recipe for cat treats.

- 1 1/2 cups rolled oats
- 1/4 cup vegetable oil
- 1/2 cup flour
- 1/2 cup tuna oil, chicken broth or beef bouillon

Preheat oven to 350° F. Mix all ingredients into a dough. Dust hands with flour and form small, 1/2-inch-thick, round "biscuits". Set on a greased cookie sheet. Bake 30 minutes, or until biscuits are slightly browned. Cool 30 minutes before serving.

How You Can Make Your Own Pet Food
www.make-stuff.com/cooking/petfood.html

> Easy instructions for dogs, cats and even budgies.

3. HEAL YOUR PETS HOLISTICALLY

If you believe herbs might be good for your health, why shouldn't your pets get the same natural treatment? Here are some suggestions.

> Herbs for Pets
> www.naturalark.com/herbpet.html
> Specific herbs for specific disorders.

POPULATION

Three Things to Do

> 1. Use birth control
> 2. Adopt
> 3. Have only one child

Details

Don't speak to me of shortage. My world is vast
And has more than enough—for no more than enough.
There is a shortage of nothing, save will and wisdom
But there is a longage of people.
—Garrett Hardin (1975)

There are currently more than six billion people on the planet. It is estimated that the world could probably naturally support two to three billion people at the most. The fact that more are currently alive is a tribute to cheap oil, which has "pushed" the planet's capabilities in order to produce enough food to feed (most of) those people. Without plentiful cheap oil, it will be impossible to support the current population.

When a population of animals or plants grows beyond its optimal size, nature has ways of bringing that population down to its appropriate size. These measures include predators, disease and natural catastrophes (catastrophic from the *population's* point-of-view; nature may feel otherwise). There is no scientific or historical reason that a human population should be an exception.

1. Use birth control

Yes, birth control measures do work. If used. With birth control, there is no reason for an unwanted — and certainly not an unexpected — child. Have a child only if you want one. Avoid the trauma of unwanted pregnancy, abortion, or an unwanted baby.

> Planned Parenthood
> www.plannedparenthood.org
> Information on health care, parenthood and birth control.

2. Adopt

Tragically, there *are* spare children in the world. If you want a child, there are ways to lovingly raise a child without actually bringing a new human into an over-crowded world. Check with your local social service agencies for information on how to start the adoption process.

> Child Welfare Information Gateway
> www.childwelfare.gov/adoption

Resources on all aspects of domestic and intercountry adoption. From the U.S. Dept. of Health and Human Services.

3. HAVE ONLY ONE CHILD

This is a hard one, but very important. The current population of China is about 1.3 billion people. It is estimated that if China had not instituted the policy of "one child per family" thirty years ago, the country's current population would be 300 million larger.

Birth control, and self-limiting the size of families, is the only way that the planet's population can be *safely* reduced. (As mentioned above, Nature has, if needed, other ways to deal with an overpopulation problem.)

Shared housing, such as cohousing (page 154), is just one of the ways that your single child can still have plenty of opportunity to play, socialize and bond with other children. You don't have to supply all of his/her playmates yourself.

World Population Awareness
www.overpopulation.org

Goal is to preserve the environment and its natural resources for the benefit of people, families, and future generations.

Dieoff
www.dieoff.org

A grim look at the future.

Sierra Club
www.sierraclub.org/population

Global population and the environment.

Tragedy of the Commons
http://en.wikipedia.org/wiki/Tragedy_of_the_commons

> Wikipedia article.
>
> **Population Connection**
> www.populationconnection.org
> Formerly *Zero Population Growth.*

CITIZEN

Three Things to Do

1. Organize your neighborhood
2. Get involved in your town
3. Lobby state and federal governments

Details

Despite what many people think, democracy was not intended to be a spectator sport. It's not simply reading about government in the paper, watching it on TV, or gossiping about Hollywood stars' latest political advocacy. It also wasn't intended to be democracy-for-hire, where we send political mercenaries a check and they do what we ask them to do.

Democracy is a *participatory* activity, or it is nothing at all. That means you get out and participate. Organize your neighborhood to accomplish neighborhood projects and to make its voice heard at city hall. Volunteer for a community group or city commission. Join or start groups to influence local policy. Run for local office or apply for appointed positions. Work with others to discuss and deal with local problems and needs.

1. ORGANIZE YOUR NEIGHBORHOOD

Democracy starts at the lowest level—your neighborhood. (See page 53.)

2. GET INVOLVED IN YOUR TOWN

Represent your neighborhood at city hall. Join with other neighborhoods to participate in city government. Get on a city commission. Run for school board or city council. Start a city-wide organization for a particular cause that needs to be addressed.

If you find that you've done everything you can for your own home and neighborhood, branch out. Take your skills and expertise to another neighborhood and see if there are ways you can assist its residents. Or check with your city government to see if they can suggest other areas where you can help.

Remember also that your community has many groups that are already organized and in place. These include religious congregations, service clubs, women's clubs, youth groups, business groups, and the like. Coordinate with them to help their members and the community as a whole implement as many of the suggestions in this book as appropriate.

3. LOBBY STATE AND FEDERAL GOVERNMENTS

There are things that state and federal governments do best because of their size and their funding. Unfortunately, they seldom do them. Peak Oil observer James Kunstler, author of *The Long Emergency*, thinks that once things really get bad in the United States, "the federal government will be lucky if it can answer the phone."

It may come to that, but in the meantime they can still be forced to do some good. Lobby your state and national elected representatives to push for action that helps communities deal with the problems resulting from Peak Oil and economic collapse. (For example, adopting the Oil Depletion Protocol – page 215, and expanding Amtrak – page 180.) Just don't be disappointed if they *don't* do anything. That's why you're reading a whole book of things you and your neighbors can do on your own—without state or federal help.

FEDERAL GOVERNMENT

Three Things to Do

> 1. Support the Oil Depletion Protocol
> 2. Turn Amtrak into a real nationwide railroad
> 3. Pass laws doubling vehicle fuel-economy

Details

1. SUPPORT THE OIL DEPLETION PROTOCOL
See page 215.

2. TURN AMTRAK INTO A REAL NATIONWIDE RAILROAD

Forget trying to make Amtrak pay its way. It needs to be heavily subsidized (as are highways and air travel) and expanded. Rail is a far more energy-efficient method of moving people and freight across land than are air and private automobile. For example, rail freight is eight times as energy-efficient as freight trucking.

Amtrak, such as it is, is the United States' only national passenger railroad system. The network currently has 22,000 miles of routes serving 500 communities (many of those connections are only by Amtrak bus) in 46 states. The only states not currently served by Amtrak are Hawaii, Alaska (which has its own system), South Dakota and Wyoming.

In 2004, more than 25 million passengers used Amtrak. By comparison, in 1916 the United States had 245,000 miles of rail, and in 1920 passenger use peaked with 1.2 billion passengers.

If we compare U.S. rail mileage per capita with other countries, we find that Australia is in first place with 2.19 km of track per 1,000 people. The United States, with about .755 km of track per 1,000 people is in 14th place, after Canada, Namibia, Kazakhstan and Hungary but a little ahead of Austria and Estonia. This does not indicate a world-class rail system, especially since the majority of our rail system is for freight only.

Friends of Amtrak
www.trainweb.org/crocon/amtrak.html

Save Amtrak
www.saveamtrak.org

National Association of Railroad Passengers
www.narprail.org

3. PASS LAWS DOUBLING VEHICLE FUEL-ECONOMY

When Henry Ford started making the Model T in 1908, that car got 25-30 miles per gallon—and it ran on either gasoline or ethanol. Ford's Model A, released in 1927, got 20-30 miles per gallon. It doesn't appear we've made much progress since then. Perhaps it's time we did. Congress should pass laws doubling vehicle fuel-economy progressively within the next five years. It's technically possible; the automakers just need to be strongly persuaded.

Want More?

- Drastically expand funding for mass transit
- Drastically expand funding for alternative energy and fuels.
- Mandate, and fund, the replacement of incandescent bulbs with compact fluorescent bulbs nationwide.

See page 29.

For Citizens

♦   Lobby the White House and members of Congress to do the above things.

♦   Throw the politicians out at the next election if they don't.

LOCAL GOVERNMENT

Three Things to Do

> 1. Change zoning to mixed use
> 2. Enforce and support water and energy conservation
> 3. Establish and support local power generation

Details

If you're involved with local government as an elected or appointed official, or a member of city or county staff, you're in a position to help make some important changes. We suggest the following as three of the most important of those changes.

1. CHANGE ZONING TO MIXED USE

Allow—and even demand—housing in commercial and light-industrial areas, and commercial in residential areas. One of the great tragedies of suburban city planning since World War II has been the ghettoization of zoning. Housing is not allowed in commercial areas; commercial uses are not allowed in residential areas. The result has been that people are forced to drive from one area to take advantage of the benefits located in the other area.

For example, few people in tract housing areas are within walking distance of grocery stores, dry cleaners, retail shops and all the other commercial services we need on a regular basis. So they, and their children, are forced to travel by car to these destinations. Fossil fuels are wasted, money is wasted, and health deteriorates due to lack of opportunity to walk. Ideally a home should never be more than a five or ten minute walk from all these shops and services—one planning concept that is supported in most urban areas.

2. ENFORCE AND SUPPORT WATER AND ENERGY CONSERVATION

Encourage and, if necessary, demand cuts in water and energy use. In support, set up a municipal plan that subsidizes such things as low-flow toilets and the replacement of incandescent bulbs with compact fluorescent bulbs.

3. ESTABLISH AND SUPPORT LOCAL POWER GENERATION

Depending on the environmental conditions, strive to generate at least some of your community's electrical power through sustainable local sources, for example wind, hydro, solar, or tidal. This could involve financially and technically assisting homes and neighborhoods, or even building municipal power plants or "energy farms".

STATE GOVERNMENT

Three Things to Do

1. Stay out of the way of local governments
2. Support alternative energy
3. Support mass transit

Details

The state is the only level of government that can handle state-wide needs. It should focus on those responsibilities and simply assist, but only when requested, county and city governments with more local matters.

1. STAY OUT OF THE WAY OF LOCAL GOVERNMENTS

They know better than the state what they need. Give them whatever they want. Focus on projects that serve regions and the entire state. If something can be done at the local level, states should give communities what they need to make it happen—with no strings. If there's something that can *only* be done at the state level, then that *is* the state's responsibility.

2. SUPPORT ALTERNATIVE ENERGY

States should give money and technical support to municipalities for their own use and/or to pass on to their local residents, but let the municipalities and their citizens decide the best way to use it. Create and fund large-scale regional and statewide sources of non-fossil fuel energy.

3. SUPPORT MASS TRANSIT

Only the state can fund and organize intrastate mass transit. Get the states to do it, whether it's railroads or transport on bays, lakes and waterways.

FURTHER READING

ALTERNATIVE HEALTH

Millions of Americans use natural and alternative methods of healing. Not all of the methods work, at least not for everyone. And there no doubt exist quacks, and frauds, and pseudo-science treatments. But so what? What we'll describe here are inexpensive or free methods of improving one's health and energy. We've tried to pick ones that are safe, but remember that any treatment, even natural herbs, can have negative effects with some people or in excess (whatever that means) quantities.

Some people will think highly of some of the suggestions, and consider other techniques to be outrageous quackery. If you see a health technique you don't like, please don't let us know. We don't care. If you don't like it, don't use it. Somebody else might find it useful.

Because, like cockroaches after a nuclear war, lawyers still exist in this new post-Peak Oil reality, we must provide the usual cover-your-ass statement declaring that we are not prescribing any form of medical treatment or advice, that we are merely providing information for educational and research purposes, and that you should always check with your physician before embarking on any health treatment program, even one performed solely by yourself on yourself.

Aromatherapy

Aromatherapy is healing using the fragrances of flowers, herbs and other plants. Even if it doesn't improve your health, it'll still smell nice.

Top 10 Essential Oils
(courtesy of the National Association of Holistic
Aromatherapy)

1. Eucalyptus – Respiratory problems
2. Ylang Ylang – Relaxation
3. Geranium – Balancing hormones in women
4. Peppermint – Headaches, muscle aches, digestive disorders
5. Lavender – Relaxation, wounds
6. Lemon – Uplifting and relaxing, wounds and infections
7. Clary Sage – Pain killer, muscular aches, relaxation, insomnia
8. Tea Tree – Anti-fungal
9. Roman Chamomile – Relaxation, muscle aches, tension
10. Rosemary – Mental and immune system stimulation

Top 10 Essential Oils
www.naha.org/top_10.htm

Aromatherapy - An A-Z [book]
Author: Patricia Davis
Essential oils, formulas and recipes.

Aromatherapy Global Online Research Archives
(AGORA)
http://users.erols.com/sisakson/pages/agoindex.htm

An international group of volunteers dedicated to
providing noncommercial aromatherapy education.

Jeanne Rose
www.jeannerose.net

Lots of articles from a recognized expert.

A World of Aromatherapy
www.aworldofaromatherapy.com

Colloidal Silver

Using silver for health is not new. It goes back centuries. More
recently, pioneers in their covered wagons crossing the
American continent in the 1800s would put a silver dollar in
the water barrel to keep the water pure.

Colloidal silver carries on the tradition. Some call it
quackery, others swear by it as an anti-bacterial and anti-viral
technique. Look into it yourself. It's somewhat expensive to
buy, but very simple and inexpensive to make, once you've
purchased a colloidal silver generator. You can find them
online for as little as $50.

From the Silver Institute.

Silver - Nature's Water Purifier
www.doulton.ca/silver.html
Benefits of colloidal silver.

Herbs

Herbs are nature's medicine. They're everywhere. More than likely, you have some in your backyard or neighborhood. If not, you can grow your own—even in your kitchen—or buy them inexpensively.

Your local garden nursery or health food store probably sells herb growing kits, or you can simply pick out the individual plants that you want at a nursery.

See also Medicinal Cooking (See page 69).

A Field Guide to Medicinal Plants and Herbs of Eastern and Central North America [book]
Author: James A. Davis.

American Botanical Council
www.herbalgramc.om
Herbal news and information.

Common Herbs
www.herbalgram.com/default.asp?c=common_herbs
29 of the most commonly used herbs in the United States.

Dr. Weil on Herbs
www.drweil.com/drw/u/id/PAG00326

From the website of Dr. Andrew Weil, a leader in integrative medicine, combing both conventional and complementary practices.

Edible and Medicinal Plants of the West [book]
Author: Gregory L. Tilford

Includes full color photographs of every plant in the book.

The Green Pharmacy [book]
Author: James A. Duke

Compendium of natural remedies from the world's foremost authority on healing herbs.

Growing and Using Herbs
www.gardenguides.com/forms/herbs.htm

For health and food - from GardenGuides.com.

HerbNet
www.herbnet.com

All-encompassing website with a huge amount of information.

Herbs for Health
www.herbsforhealth.com

Magazine with lots of online articles and information.

Identifying and Harvesting Edible and Medicinal Plants
[book]
Author: Steve Brill

In wild (and not so wild) places.

Making Cayenne Tincture
www.curezone.com/schulze/tinctures.asp

Instructions for making an herbal tincture. Use the same method for other herbs.

Natural Herbs Guide
www.naturalherbsguide.com

Herbs, remedies, medicines and supplements.

Stalking the Healthful Herbs [book]
Author: Euell Gibbons

By the author of Stalking the Wild Asparagus.

A few all-purpose, anti-bacterial, anti-viral remedies are good to have around the home. These recipes are claimed to fit that need.

Four Thieves

Legend has it that a group of four thieves looted jewelry and other valuables from countless victims (some of whom they dispatched themselves) of the Great Plague in Europe, and that they never got the bubonic plague themselves due to a secret compound that they rubbed on their bodies.

This site gives four different recipes for this legendary preventive medicine.

Four Thieves
www.kitchendoctor.com/articles/four_thieves.html

Master Tonic - A Natural Antibiotic
www.nspforum.com/index.cgi?read=15534
This is another all-purpose treatment. It's Dr. Richard Schulze's adaption of Dr. John Christopher's original anti-plague tonic.

Make Your Own Herbal Tinctures
www.kcweb.com/herb/tincture.htm
Make your own from any fresh herb.

Light

Perhaps it's time to let go of our paranoia about being attacked by sunlight, and start to recognize how healthy the source of all life energy is. Just 10-15 minutes out in the sun each day can help the body get healthier.

For people who experience Seasonally Affected Disorder (SAD), it's essential that they get exposure to sunshine during the winter months.

If there's no sun outside, install full-spectrum bulbs in your home. These bulbs provide the full range of light that traditional incandescent—and even most fluorescent—bulbs can't provide. They're better for your eyesight and overall for your health.

Light: Medicine of the Future [book]
Author: Jacob Liberman

How we can use it to heal ourselves.

Ott Lite
www.ottlite.com

Full spectrum lights were pioneered by John Ott, a former cinematographer for Walt Disney.

Guilt-Free Sunbathing
www.findarticles.com/p/articles/mi_m0NAH/is_4_32/ai_85
174712

Why the sun can be good for your health.

The Benefits of Sunlight
www.whatreallyworks.co.uk/start/articles.asp?article_ID=
451

Information on ailments that can be prevented or relieved.

Solar Healing Center
www.solarhealing.com

Official site for Hira Ratan Manek, the guru (literally) of sungazing.

Sungazing
www.sungazing.com

Personal experiences and instructions for practicing sungazing.

Sungazing Forum
http://health.groups.yahoo.com/group/sungazing

Yahoo discussion group.

Massage

Massage doesn't just feel good, it's healthy. There are many books you can get to teach you various massage techniques. Many communities also offer classes in massage through adult school and local community colleges.

If you'd rather learn online, you'll find step-by-step instructions at some of the sites below.

Erotic Massage
www.sexuality.org/erotmass.html

Instructions for giving and receiving erotic massage. It's more than just healthy.

Free Massage
www.massagefree.com

Gives detailed massage instruction in more than 50 free videos. Focus is sexual but the techniques themselves are actually standard.

Learn the Art of Self Massage
www.rd.com/content/openContent.do?contentId=16051

Effective, simple techniques for home or office.

Massage
www.mckinley.uiuc.edu/Handouts/Massage/massage.htm l

Instructions from the health center at the Univ. of Illinois, Urbana-Champaign.

Qigong Self Massage
www.mnwelldir.org/docs/qigong/qgsm.htm

Detailed instructions for Chinese method of self-massage.

Self-Massage Techniques
www.kreinik.com/articles/news.php?newsid=32

Good instructions covering almost the entire body—with illustrations.

Reflexology / Acupressure

Reflexology is a form of acupressure in which healing is done by pressing and stimulating specific zones of the hands and feet. Corresponding parts of the body are energized by this pressure. For example, the end of the thumb "connects" with the head, brain and sinuses; while the center of the bottom of the big toe is associated with the pituitary gland.

Acupressure Institute
www.acupressure.com

Good articles from the author of Acupressure's Potent Points.

Foot Reflexology Maps
www.dorlingkindersley-
uk.co.uk/static/cs/uk/11/features/reflexology/extract.html

Online interactive maps of the reflexology zones on the hands and feet—an outstanding tool. Also downloadable foot and hand charts.

How to Apply Pressure to Points When Doing Shiatsu
www.ehow.com/how_8591_apply-pressure-points.html

Shiatsu is Japanese acupressure.

Reflexology Massage Instructions
http://groups.msn.com/AlternativesToPainandDisease/refl
exologyinstructionspg1.msnw

Charts and instructions for doing reflexology.

What is Reflexology?
www.reflexology-research.com/whatis.htm

Frequently asked questions.

Relaxation and Meditation

There is nothing more relaxing than relaxing. While there are lots of things people can do for "relaxation", whether it might be watch TV, play golf, or knit, the best relaxation most of us can do is to simply stop and do nothing. True relaxation removes the tension from every part of the body. It rejuvenates the body, calms and invigorates the mind, and just plain feels good. Add meditation, and you'll experience a feeling of energetic peace and contentment that can keep you going the entire day.

Meditation needn't take a lot of time. One of the best-known techniques—Transcendental Meditation—suggests two twenty-minute sessions a day. Even if you can initially only fit in ten minutes a day with whichever technique you use, you'll find it rewarding. And with time it's easier to fit in longer stretches.

All religions use forms of meditation, whether they call it that or not, but you don't have to join a religion in order to meditate. No matter where you live you can find low-cost or even free instruction in a variety of meditation techniques. People meditate sitting down, lying down, standing up, or even walking. It can be as simple as focusing on your breath, or as complicated as combining visualization, breathing, chanting and movements all at the same time. We suggest you start with the simple. It begins with breathing.

Breathe Better

Breathing properly is something few of us do, although it's essential to good health.

"What?", you say. "I know how to breathe." Well, yes. But most of us do shallow breathing using only the top of our lungs, never using our full lung capacity. For optimum air and oxygen intake, as well as taking in the energy referred to as "prana" (in India) or "chi" (in China), we should breathe deeply from the bottom of our lungs.

Here's an exercise that will help you improve your breathing. If you do the exercise on a regular basis, your day-to-day breathing will improve.

Deep Breathing

Deep Breathing allows us to use our entire lungs, providing more oxygen to our bodies, and energizing and rejuvenating every organ and cell in our bodies. It is probably the most effective and beneficial method of relaxation we've seen.

1. Lie on your back.
2. Slowly relax your body, starting with your feet and moving through every part of your body until you have reached—and relaxed—your face and scalp.
3. Do a quick check to see if you've missed any place. If so, relax it.
4. Slowly begin to inhale, first filling your lower belly, then your stomach area, and then your chest and the top of your lungs almost up to your shoulders. Hold for a second or two, then begin to exhale. Empty the very bottom of your lungs first, then the middle, then finally the top.
5. Continue this breathing for four or five minutes. Don't force your breathing; it's not a contest to see how much air you can take in. Just do it in a relaxed, peaceful manner.
6. After a while, imagine that you are resting on a warm, gentle ocean. The sun is shining peacefully on your body. Imagine that you rise on the gentle swells of the water as you inhale, and that you slowly descend as you exhale.
7. Continue this relaxing breathing as long as you wish.

Anuloma Viloma Pranayama
www.abc-of-yoga.com/pranayama/basic/viloma.asp

An ancient Indian breathing exercise using alternate nostril breathing.

Relax More

Here's one of the most popular relaxation exercises. A full-body relaxation while lying down that goes deeper than the relaxation exercise above. It's particularly useful if you're having a hard time trying to get to sleep at night.

Progressive Relaxation

This exercise is most effective when you tape record the instructions in advance, preferably in your own voice. This way you don't have to concentrate on remembering the instructions.

We'll give you the instructions here. You tape record them, with a short pause after each sentence to allow yourself time to actually do the sensing and relaxing.

1. Lie on your back, wiggle around until you're as comfortable as you can get, close your eyes, and begin to listen to the tape.
2. Feel your feet. Feel the weight of your feet. Feel your feet relax and sink into the bed.
3. Feel your lower legs. Feel the weight of your lower legs. Feel your lower legs relax and sink into the bed.
4. Feel your knees. Feel the weight of your knees. Feel your knees relax and sink into the bed.
5. Feel your upper legs. Feel the weight of your upper legs. Feel your upper legs relax and sink into the bed.
6. Feel your hands. Feel the weight of your hands. Feel your hands relax and sink into the bed.
7. Feel your lower arms. Feel the weight of your lower arms. Feel your lower arms relax and sink into the bed.
8. Feel your elbows. Feel the weight of your elbows. Feel your elbows relax and sink into the bed.
9. Feel your upper arms. Feel the weight of your upper arms. Feel your upper arms relax and sink into the bed.
10. Feel your buttocks. Feel the weight of your buttocks. Feel your buttocks relax and sink into the bed.
11. Feel your back. Feel the weight of your back. Feel

your back relax and sink into the bed.

12. Feel your pelvic and belly area. Feel the weight of your pelvic and belly area. Feel your pelvic and belly area relax and sink into the bed.

13. Feel your chest. Feel the weight of your chest. Feel your chest relax and sink into the bed.

14. Feel your shoulders. Feel the weight of your shoulders. Feel your shoulders relax and sink into the bed.

15. Feel your neck, both front and back. Feel the weight of your neck. Feel your neck relax and sink into the bed.

16. Feel your skull. Feel the weight of your skull. Feel your skull relax and sink into the bed.

17. Feel your mouth. Feel any tension in your mouth. Feel your mouth relax and any tension slide off into the bed.

18. Feel your eyes. Feel any tension in your eyes. Feel your eyes relax and any tension slide off into the bed.

19. Feel your entire face. Feel any tension in your face. Feel your face relax and let any tension slide off into the bed.

20. Mentally scan your body. If you find any place that's still tense, relax it and let it sink into the bed.

Meditation

Meditation is simply focusing the mind to the point of stillness. It's usually done in a repetitive, even ritualistic, manner. It is variously called meditation, contemplation, and even prayer. Meditation can involve complex visualizations or a simple focusing on the breath or on a sound repeated over and over again.

We provide information on websites that will give you instructions for a very wide variety of meditation techniques.

Holistic Online
http://1stholistic.com/Meditation/hol_meditation.htm

Extensive information on meditation and its physical, psychological and spiritual benefits. Also includes instructions for a number of meditations from a variety of disciplines.

Learning Meditation
www.learningmeditation.com/room.htm

More than 15 guided meditations that you can listen to on your computer.

Meditation
http://en.wikipedia.org/wiki/Meditation

From Wikipedia.

Meditation Center
www.meditationcenter.com

Directions for a variety of meditation techniques.

Meditation Station
www.meditationsociety.com

Website of the Meditation Society of America. Many different techniques.

Sound

Sound is one of the most sure-fire methods for relaxation— and meditation.

Music

Listening to soothing music works for most people, but the type of music they find most relaxing varies. We suggest some of the quieter classical recordings. "New Age" music is popular, though some people find it a bit too saccharine. Many people prefer just listening to nature sounds, the breaking waves of the ocean, wind rippling through the trees, or the bubbling sound of a mountain brook.

Mantra

An ancient method of using sound for relaxation and meditation is "mantram". Mantra (plural) are sounds repeated over and over again (chanted). They are generally in one of the more ancient languages, such as Sanskrit, Hebrew or Arabic. Perhaps the best known one is "Om", pronounced "ommmmmmmm".

Mantra
http://en.wikipedia.org/wiki/Mantra
From Wikipedia.

Singing

One of the best, and most personal, methods of relaxation is singing. Your own singing. Set aside fifteen minutes a day to sing. If you're shy about it, do it in the car. Or at home when you're alone. Cars and showers, of course, give us great acoustics and make us sound much better. (At least we *think* we sound better.)

Sing your favorite songs, whichever you enjoy the best. If you don't know all the words, fake it. If you want to know all the words, go on the Internet. You'll find the lyrics for most any song you can think of.

LyricsDownload.com
www.lyricsdownload.com
Free lyrics database.

One of the nicest things about this method of relaxation is you could do it almost anywhere (well, almost) and it costs absolutely nothing.

Vision

Methods have been developed for improving your eyesight with exercise, rather than lenses or surgery. The most popular are called the "Bates Exercises", named after the founder. The

Bates Method is based on the theory that the mind controls the eyes, and that if the mind learns to relax, the eyes will also.

At the least, the exercises will relax your eyes (and mind), keeping them healthier and lessen the risk of headaches. At the most, as thousands have reported, they can actually improve your vision and maybe even help you get along without glasses.

Bates Association for Vision Education
www.seeing.org
Association of Bates Method teachers.

Imagination Blindness
www.iblindness.org
A website with extensive information on the use of the Bates method.

Perfect Sight Without Glasses
www.i-see.org/perfect_sight
The original book by Dr. William Bates, published in 1920, and available online free in its entirety.

Vision sites on the Web
www.seeing.org/links.htm
Links to Bates practitioners and other methods of vision enhancement.

Yoga

Yoga is a body of ancient Indian spiritual practices. The branch of yoga most people are aware of is Hatha Yoga, which is designed to purify and strengthen the body, improving it as a vehicle for meditation. One of the many benefits of yoga is that you don't have to buy anything to do it.

Alternative Medicine Man
www.yogacards.com/Alternative-Medicine-Man.html

Yoga sequences for specific ailments.

Yoga Basics
www.yogabasics.com/asana

Complete instructions for more than 100 yoga.

Yoga Journal
www.yogajournal.com

The classic magazine with a huge amount of online information.

BARTER ITEMS

Here's a list of barter items, in no particular order, that circulated during Y2K days. They're items that are historically, or at least likely, to be difficult to find but very much needed during hard times. If you'd like to stock up on stuff, whether for your own use or for trade, you might find this list useful.

We're sure you can think of other things to add to the list appropriate to your specific needs and where you live.

1. Generators
2. Water filters/purifiers
3. Portable toilets
4. Seasoned firewood
5. Lamp oil, wicks, lamps
6. Coleman fuel
7. Hand-operated can openers, hand egg beaters, whisks
8. Honey/syrups/white, brown sugars, and other sweeteners

9. Rice, beans, wheat
10. Vegetable oil (for cooking)
11. Charcoal and lighter fluid
12. Water containers
13. Grain grinder (non-electric)
14. Propane dylinders
15. Lantern mantles: Aladdin, Coleman, etc.
16. Baby supplies: diapers/formula/ointments/aspirin, etc
17. Washboards, mop bucket w/wringer (for laundry)
18. Cookstoves (propane, Coleman, kerosene)
19. Vitamins
20. Feminine hygiene/haircare/skin products
21. Thermal underwear
22. Bow saws, axes and hatchets and wedges (also, honing oil)
23. Aluminum foil
24. Gasoline containers (plastic or metal)
25. Garbage bags
26. Toilet paper, facial tissues, paper towels
27. Milk - powdered and condensed
28. Garden seeds (non-hybrid)
29. Clothes pins/line/hangers
30. Canned tuna
31. Fire extinguishers
32. First aid kits
33. Batteries
34. Garlic, spices, vinegar, baking supplies
35. Flour, yeast and salt
36. Matches (preferably strike-anywhere)
37. Writing paper/pads/pencils/solar calculators
38. Insulated ice chests
39. Workboots, belts, jeans and durable shirts
40. Flashlights, lightsticks, lanterns
41. Journals, diaries and scrapbooks
42. Garbage cans, plastic

43. Men's hygiene: shampoo, toothbrush/paste, Mouthwash/floss, nail clippers,etc
44. Cast iron cookware
45. Fishing supplies/tools
46. Mosquito coils/repellent sprays/creams
47. Duct tape
48. Tarps/stakes/twine/nails/rope/spikes
49. Candles
50. Laundry detergent (liquid)
51. Backpacks and duffle bags
52. Garden tools and supplies
53. Scissors, fabrics and sewing supplies
54. Canned fruits, vegetables, soups, stews, etc.
55. Bleach (plain, not scented: 4 to 6% sodium hypochlorite)
56. Canning supplies (Jars/lids/wax)
57. Knives and sharpening tools: files, stones, steel
58. Bicycles - tires/tubes/pumps/chains, etc.
59. Sleeping bags and blankets/pillows/mats
60. Carbon monoxide alarm (battery powered)
61. Board games , cards, dice
62. d-Con rat poison, Mouse Prufe II, roach killer
63. Mousetraps, ant traps, cockroach magnets
64. Paper plates/cups/utensils
65. Baby wipes, oils, waterless and anti-bacterial soap
66. Rain gear, rubberized boots, etc.
67. Shaving supplies (razors and creams, talc, after-shave)
68. Hand pumps and siphons (for water and for fuels)
69. Soy sauce, vinegar, boullion/gravy/soup base
70. Reading glasses
71. Chocolate/cocoa/tang/punch (water enhancers)
72. Woolen clothing, scarves/earmuffs/mittens
73. Boy Scout handbook
74. Graham crackers, saltines, pretzels, trail mix/Jerky
75. Popcorn, peanut butter, nuts

76. Socks, underwear, t-shirts, etc.
77. Lumber
78. Wagons and carts
79. Cots and inflatable mattresses
80. Gloves: work/warming/gardening, etc.
81. Lantern hangers
82. Screen patches, glue, nails, screws, nuts and bolts
83. Teas
84. Coffee
85. Cigarettes
86. Beer/wine/liquors
87. Paraffin wax
88. Glue, nails, nuts, bolts, screws, etc.
89. Chewing gum/candies
90. Atomizers (for cooling/bathing)
91. Hats and cotton neckerchiefs
92. Goats/chickens
93. Guns, ammunition, pepper spray, knives, clubs, bats and slingshots (depends on how warlike you feel)

CUBA

Cuba has already been through economic collapse as a result of the shortage of energy resources, and there may be much we can learn from that country. Their own "Peak Oil" happened after the Soviet Union collapsed in 1989 and Cuba lost its primary sugar market and the source of most of its petroleum and other fossil fuels and raw materials.

The Cubans rose to the occasion, and today Cuba is a model of sustainability for the rest of the world. But it required much sacrifice — and it was done at a time when the world economy was still functioning and ample energy resources were available in other countries.

Cuba euphemistically refers to this difficult time—which it is still coming out of—as its "Special Period" (*Período especial*). The Cubans responded to shortages in ways which we may all soon have to consider doing.

Prior To The Crash

Prior to the collapse of the USSR, Cuba received huge support from the Soviet Bloc countries. Cuban sugar was purchased at five times the world price, and the Soviet countries provided Cuba with 98% of its petroleum. Cuba received so much petroleum that it was able to re-sell some of that oil on the world market in order to gain hard currency.

As a result of this heavily-subsidized fossil fuels supply, Cuba was heavily dependent on gasoline and natural gas for transportation and agriculture. It used tractors—more than 20,000 of them—on its large, industrial, state-owned farms (one agronomist said that Cuba had more tractors per acre than California). Like the United States, Cuba heavily treated its agriculture with pesticides, insecticides, and herbicides.

The country exported trade crops such as tobacco, sugar, coffee and citrus fruits and imported 66% of its basic staples. It also imported 86% of all raw materials, and 80% of machinery and spare parts, most of this from Soviet Bloc countries.

Results Of The Crash

When trade with the USSR disappeared, Cuba lost 53% of its petroleum imports and 85% of its trade economy. It also lost 77% of its imported fertilizer, dropping from 1.3 million tons per year to 300,000 tons. Its imports of animal feed dropped 70%, and the dollar value of its pesticide imports fell 63%.

Caloric Intake

Prior to the Soviet collapse, food imports constituted 66% of the food consumed by Cubans. Those imports dropped by 50%. The country's own food production dropped by 45%.

The daily per capita caloric intake decreased 36%, protein intake dropped 40%, and dietary fats fell 65%. Most Cuban adults lost an average of 20-30 pounds in weight.

Thanks to strong government programs, the results of food shortages were not as bad as they could have been. Children, the elderly, and pregnant and lactating mothers were ensured healthy levels of caloric intake, and a ration card system guaranteed everyone a basic, minimum level of food. In 1993, the average daily intake was 1863 calories and 46 grams of protein, considerably below the World Health Organizations' recommended 2400 calories and 72 grams of protein; today it has returned to an average of more than 3300 calories and 82 grams.

Today's Cuban Agriculture

Prior to the Special Period, there were already a number of scientists and agricultural experts who had been urging the country to move toward more organic and sustainable methods of farming in order to combat soil erosion and mineral depletion. When oil imports collapsed, they moved into action. The result is perhaps the most organically oriented and self-sustainable (but not totally sustainable) country in the world.

Cuba today uses oxen instead of tractors wherever possible. "Animal traction" saves fuel, avoids soil compaction and turns the soil to a healthier level. The number of teams of oxen in the country increased from 50,000 to 400,000. As a side benefit, there was a new cottage industry in harness shops, and the number of blacksmiths quadrupled.

Cuba was formerly dependent on petroleum-based pesticides and natural gas-based fertilizers. Today the country mainly uses *biofertilizers* such as manure, compost, vermiculture (earthworms—see page 81), nitrogen-rich cover crops (green manure) and crop rotation; and *biopesticides* such as microbes, natural biological predators, resistant varieties of plants, and complementary crops.

Cuba has become perhaps the most organically farmed

country in the world (some crops, such as rice and tobacco, continue to require chemical-based fertilizers and pesticides). This was not out of choice, but out of necessity. Cuba today uses less than 1/20th of the amount of pesticides that it used before 1989. The Cubans have decreased, by necessity, the amount of meat in their diet and increased the amount of fruits and vegetables. The health of the people (for example, a 25% decline in heart disease) and the coffers of the country's government benefit greatly from the change.

Community/Urban Gardens

The Cuban government and its people created community gardens in cities and towns throughout Cuba. (Actually the people started it first; the government then got on the bandwagon and supported them.) The larger state-supported and cooperative organic farms are called *organopónicos;* smaller neighborhood and patio gardens are *huertos.*

The outskirts of cities provide still more space for gardens, and there are even gardens attached to factories and office buildings, where the workers are able to produce food for lunchtime meals and to take home. All of these types and sizes of gardens increased the country's food production, and avoided the need for fuel to transport food from the countryside into the cities.

Havana has a population of about two million people, and its gardens currently produce more than 50% of the vegetables its residents consume. Urban gardens throughout the country produce 60% of all the vegetables consumed in the country. Community gardens produce, by law, only organic food. Most goes to the people who work in the gardens. Surplus is donated to schools, senior centers, and other community programs, or is sold in farmers markets.

Many farms and gardens use the natural farming methods of Permaculture (see page 78) and Biodynamics (see page 77).

Transportation, Energy, And Medicine

With the dramatic drop in vehicle fuel, the government turned to shared transportation. It encourages and supports the use of buses, carpooling, and hitchhiking. (All government vehicles are required to pick up hitchhikers.)

Cubans are walking and biking much more. Cuba bought 1.2 million bicycles from China (this in a country of less than 12 million people), and produced another half million of their own. However, Cubans never really fell in love with the concept of bicycling—maybe because most of the bikes were heavy and only one-speed—and bicycles are seen more often these days in rural areas and smaller towns than in cities such as Havana.

Cuba provides free health care to all. Since the beginning of the Special Period, the focus has been on *preventive* medical care in order to avoid the higher costs of treatment. Cuba is a world leader in biotechnology, and has developed new vaccines and treatments for such diseases as malaria, hepatitis and dengue fever. The country has also turned more and more to "green medicine", using herbs and other natural remedies, and increasing the use of such techniques as acupuncture, aromatherapy, yoga, tai chi, and other health-promoting and non-pharmaceutical procedures.

Cubans per capita use five to six percent of the energy consumed by residents of the United States. To conserve fossil fuels and to generate energy close to the actual users of that energy, Cuba has moved toward small-scale renewable energy production. It has begun using photovoltaic panels to generate solar energy throughout the country. Solar currently provides energy to more than 2,000 schools and hundreds of hospitals.

In a current energy program the government is distributing more than nine million free compact fluorescent light bulbs (see page 29) to Cuban households. In further efforts to reduce energy consumption the government is distributing 3.5 million electric rice cookers, 2.5 million electric cooking pots, 2.3 million water heaters, and 250,000 energy efficient refrigerators, all of which will use much less energy than the appliances currently being used.

Embargo

Cuba has been under a U.S. embargo since 1962 and these days it is almost impossible for the average American to visit Cuba. Cuba welcomes American visitors, as it welcomes visitors from all over the world. Americans are the only people not allowed by their own government to visit the island nation.

If you're an American and would like to see Cuba for yourself to learn first-hand how they dealt with Peak Oil, you might want to support those groups working to end the embargo.

History of the Blockade
http://www.globalexchange.org/countries/americas/cuba/background/history.html

Information from Global Exchange.

Cuba Program
http://www.ciponline.org/cuba/

Center for International Policy.

The Power of Community
www.globalpublicmedia.com/articles/657

How Cuba Survived Peak Oil (article).

The Power of Community
www.communitysolution.org/cuba.html

How Cuba survived Peak Oil (DVD).

Cuba's transition from an industrial petroleum-based society to a sustainable society, as a result of their loss of petroleum when their source the Soviet Union collapsed.

Cuba - A Hope
www.fromthewilderness.com/free/ww3/
120103_korea_2.html

How Cubans successfully responded to their own
equivalent of Peak Oil.

Food Security in Cuba
http://www.monthlyreview.org/0104koont.htm

Excellent, comprehensive article covering large-scale and
community agriculture.

Cuba and Venezuela Lead Global Organic Revolution
www.organicconsumers.org/corp/cubavenez20205.cfm

Diversified agricultural production in the two countries.

Cuba Diet, The
www.harpers.org/TheCubaDiet.html

How Cuba dealt with food shortages. Can industrial
countries do it too?

Cuba's Organic Food Revolution Flourishing
www.organicconsumers.org/organic/cuba_organic_food.cf
m

More on Cuban agriculture.

Cuba's Jewel of Tropical Medicine
www.globalexchange.org/countries/americas/cuba/sustain
able/natTradMedicine/1496.html

Article on the Pedro Kourí Institute of Tropical Medicine.

Green Medicine
http://archive.salon.com/health/feature/2000/01/26/medicin
e_verde

How Cuba is integrating natural remedies into its public
health care - Salon article.

Latin American School of Medicine
www.elacm.sld.cu

Official Spanish language site of *Escuela
Latinoamericana de Medicina* that offers free medical

training for students from throughout the Americas and other regions (including poor areas of the United States).

Information on the school in English)
http://cubamigo.org/juliancindy/medschool2.html

Scholarship Program
www.ifconews.org/MedicalSchool/main.htm

including scholarships for U.S. students.

Cuba
www.beyondpeak.com/cuba-beyondpeak.html

A more extensive list of links on Cuba's "Peak Oil" response.

The Cuba Page
www.boomersabroad.com/cuba.html

A metadirectory of information and links on all aspects of Cuba.

HEMP

Hemp is not marijuana. It is a "cousin" of marijuana, with very different qualities.

Similarities

- ◆ Hemp and marijuana are varieties of the Cannabis sativa plant.
- ◆ They're both illegal to grow in the United States, although hemp *products* are legal.
- ◆ Both have similar aromas when in bloom.

Differences

- ♦ Smoking marijuana makes the user high. Smoking hemp doesn't. (Hemp contains less than 1 percent of the active ingredient THC, marijuana has 10 to 20 percent.)
- ♦ Marijuana plants tend to be short and bushy, hemp can be as high as 25 feet tall.
- ♦ Marijuana can be used to smoke or eat. Hemp can be used to produce more than 25,000 products.

Hemp as Food

As HempFood.ca (www.hempfood.ca) says:

"There is an answer to world hunger, to over-fertilization, to global warming, to the overuse of fossil fuels, to the obesity issue, to poor health and weak cardiovascular systems. The answer is hemp and hemp seed food."

Hemp seed is a highly nutritious source of protein and essential fatty oils. The protein in hemp seed closely resembles protein as it is found in human blood. It is extremely easy to digest. The essential fatty acids that hemp seed provides contain almost no saturated fat. As a diet supplement, they can reduce the risk of heart disease. One handful of hemp seed per day will supply adequate protein and essential oils for an adult.

Hemp requires little fertilizer and water, and grows well almost everywhere. It also resists pests, so it uses few if any pesticides. Hemp puts down deep roots, which prevents topsoil erosion, and when the leaves drop off the hemp plant, minerals and nitrogen are returned to the soil. Hemp has been grown on the same soil for twenty years in a row without any noticeable depletion of the soil.

Hemp seeds are an excellent source of both soluble fiber and insoluble fiber. They are an excellent way to gently cleanse the intestinal system and are a welcome addition of heavy fiber to anyone's diet. Hemp can also be a valuable, and highly productive, source of biodiesel fuel.

But nutrition is just one of the many benefits of hemp.

Other Uses of Hemp

Hemp grows quickly, maturing in less than four months. It grows in most climates and naturally resists disease; therefore pesticides are not necessary. Hemp removes weeds from the area where it grows and produces up to four times the amount of mass as the same acreage of trees. Once harvested, the soil remains fertile.

Hemp has strong fibers, the paper it produces is naturally acid-free (no "yellow journalism" since it does not turn yellow or brittle with age). Hemp paper can be recycled more than twice as many times as wood pulp paper, and it can be bleached with environmentally safe hydrogen peroxide rather than the toxic chlorine bleach (which results in dioxin) used to make new wood pulp paper.

The pulp can be used to make fuel or produce acid-free paper. It makes an excellent cloth, and its fiber can be used to make rope or twine. The seed is high in protein and essential fatty acids and can be eaten by both humans and animals. The oil from the seed can be used as a base for paints and varnishes. A tincture of the resin in the blossoms and leaves can be used as medicine, the pulp can be burned as fuel or processed to produce methanol or ethanol, it can be used in the construction industry to make composite board, and it can be used to make lubricants or plastics,

Hemp can be used to treat multiple sclerosis, cancer, AIDS (and AIDS treatment side effects), glaucoma, depression, epilepsy, migraine headaches, asthma, and severe pain, among many other ailments and diseases.

Hemp has been used successfully in the world for more than 3,000 years.

Hemp
www.beyondpeak.com/hemp-beyondpeak.html

Further information, links, and why hemp was criminalized.

Dr. Dave's Hemp Archives
www.gametec.com/hemp

Extensive information from a long time scientific researcher into plant breeding and genetics.

History and Benefits of Hemp
http://www.sdearthtimes.com/et0199/et0199s11.html

From Earth Times.

The Emperor Wears No Clothes
www.jackherer.com/chapters.html

Free online version of the classic book (more than 600,000 sold).

Hemp Industries Association
www.thehia.org
Non-profit trade group representing hemp companies, researchers and supporters.

OIL DEPLETION PROTOCOL

About the Protocol

The Oil Depletion Protocol is a proposed international agreement that will enable nations of the world to cooperatively reduce their dependence on oil—thus, it is hoped, avoiding wars over petroleum, terrorism and economic collapse. It was proposed by Dr. Colin Campbell, a prominent petroleum geologist and founder in 2002 of the Association for the Study of Peak Oil and Gas (ASPO).

The reality is that the countries of the world are going to reduce their use of petroleum *whether they like it or not.* Geology and oil availability will bring this about. Without mutual cooperation, the declining levels of oil production will lead to wars over oil, terrorist attacks on oil production facilities and pipelines, and worldwide economic collapse, as countries struggle individually and competitively to deal with the problems of Peak Oil.

The Oil Depletion Protocol provides a way for countries to work together to lessen the problems resulting from Peak Oil. Because the decrease in oil production will be known in advance, long-range planning can be done and pricing can be stabilized. At the same time, as use of fossil fuels decreases in a mutually-planned process, the Oil Depletion Protocol works hand-in-hand with the Kyoto Protocol to reduce carbon dioxide emissions, decreasing further contributions to global warming.

Oil-importing countries would agree to reduce their imports each year by the "World Depletion Rate", which is currently about 2.6 percent. Oil-exporting countries, who are already decreasing their production anyway (or soon will be), would produce less oil each year, at their "National Depletion Rate". Many exporters will actually be decreasing their exports at a faster rate because of an already-existing higher depletion rate.

The National Depletion Rate is found by determining the amount yet to be produced (known reserves plus likely amount still to be found) and dividing that by the yearly amount currently being extracted.

Text of the Oil Depletion Protocol

(As drafted by Dr. Colin J. Campbell)

WHEREAS the passage of history has recorded an increasing pace of change, such that the demand for energy has grown rapidly in parallel with the world population over the past two hundred years since the Industrial Revolution;

WHEREAS the energy supply required by the population has come mainly from coal and petroleum, such resources having been formed but rarely in the geological past and being inevitably subject to depletion;

WHEREAS oil provides ninety percent of transport fuel, is essential to trade, and plays a critical role in the agriculture needed to feed the expanding population;

WHEREAS oil is unevenly distributed on the Planet for well-understood geological reasons, with much being concentrated in five countries bordering the Persian Gulf;

WHEREAS all the major productive provinces of the World have been identified with the help of advanced technology and growing geological knowledge, it being now evident that discovery reached a peak in the 1960s, despite technological progress and a diligent search;

WHEREAS the past peak of discovery inevitably leads to a corresponding peak in production during the first decade of the 21st Century, assuming no radical decline in demand;

WHEREAS the onset of the decline of this critical resource affects all aspects of modern life, such having grave political and geopolitical implications;

WHEREAS it is expedient to plan an orderly transition to the new World environment of reduced energy supply, making early provisions to avoid the waste of energy, stimulate the entry of substitute energies, and extend the life of the remaining oil;

WHEREAS it is desirable to meet the challenges so arising in a co-operative and equitable manner, such to address related climate change concerns, economic and financial stability, and the threats of conflicts for access to critical resources.

NOW IT IS PROPOSED THAT

A convention of nations shall be called to consider the issue with a view to agreeing an Accord with the following objectives:

- to avoid profiteering from shortage, such that oil prices may remain in reasonable relationship with production cost;
- to allow poor countries to afford their imports;
- to avoid destabilizing financial flows arising from excessive oil prices;
- to encourage consumers to avoid waste;
- to stimulate the development of alternative energies.

Such an Accord shall have the following outline provisions:
- The world and every nation shall aim to reduce oil consumption by at least the world depletion rate.
- No country shall produce oil at above its present depletion rate.
- No country shall import at above the world depletion rate.
- The depletion rate is defined as annual production as a percent of what is left (reserves plus yet-to-find).
- The preceding provisions refer to regular conventional oil — which category excludes heavy oils with cut-off of 17.5 API, deepwater oil with a cut-off of 500 meters, polar oil, gas liquids from gas fields, tar sands, oil shale, oil from coal, biofuels such as ethanol, etc.

Detailed provisions shall cover the definition of the several categories of oil, exemptions and qualifications, and the scientific procedures for the estimation of Depletion Rate.

The signatory countries shall cooperate in providing information on their reserves, allowing full technical audit, such that the Depletion Rate may be accurately determined.

The signatory countries shall have the right to appeal their assessed Depletion Rate in the event of changed circumstances.

> (Note: the Oil Depletion Protocol has elsewhere been published as "The Rimini Protocol" and "The Uppsala Protocol." All of these documents are essentially identical.)

What you can do

1. LOBBY THE WHITE HOUSE

Phone, write, email, and use any contacts you have that can

influence the administration.

2. LOBBY CONGRESS

Phone, write, email, and use any contacts you have that can influence your Representative or Senators. Contact both their local and Washington, D.C. offices.

3. WRITE LETTERS TO THE EDITOR

Write about the Protocol and the dangers it seeks to avoid. Encourage members of your community to contact politicians as well.

Colin Campbell
www.hubbertpeak.com/campbell
Dr. Campbell's website.

ASPO
www.peakoil.net
Association for the Study of Peak Oil.

Oil Depletion Protocol
www.oildepletionprotocol.org
Book and website by Richard Heinberg with full information on the protocol, its significance, and how you can help make it happen.

SECESSION

Secession is an extreme method of bringing government and decision-making closer to the people who are affected by those decisions. It has become clear to many people that modern countries, particularly those with the land area and population of a country like the United States, are too big to govern properly. Combine that with the need to devolve government down to lower levels, and the solution is to break the larger countries up into smaller, more manageable countries; or to break large states of a country into smaller states. Recognizing the rights of those areas to secede from a national or state entity is a first step.

> **American Secession Project**
> **www.secessionist.us**
> **Dedicated to placing secession in the mainstream of political thought as a viable solution to contemporary problems.**
>
> **Secession Net**
> **www.secession.net**
> **Promoting the right to secede.**

Here are a few examples of secession efforts in the United States.

Cascadia

A proposal to form a country composed of the Canadian province of British Columbia and the American states of Washington and Oregon.

> **Republic of Cascadia**
> **www.zapatopi.net/cascadia/**
> **Home page for the Republic of Cascadia.**

Confederate States of America

Although the southern states of the United States failed in

their last attempt (variously named "The War of Northern Aggression" and "The Recent Unpleasantness") to voluntarily secede from the Union that they had voluntarily joined, we encourage them to consider it again.

Hawaii

Many Hawaiians advocate the return of total control of the islands to the Hawaiian people. The Americans overthrew Queen Lili'uokalani in 1893 and have militarily occupied the islands since 1898.

> **Hawaiian Independence**
> **www.hawaii-nation.org**
> **Covers all aspects of Hawaiian self-determination.**

Jefferson

Dedicated to the proposed State of Jefferson, formed from the counties of Southern Oregon and Northern California.

> **Jefferson State**
> **www.jeffersonstate.com**
> **Jefferson State website.**

Vermont

A peaceful, democratic, grassroots solidarity movement committed to the return of Vermont to its rightful status as an independent republic as was the case in 1791 and to support Vermont's future development as a separate, sustainable nation-state.

Second Vermont Republic
www.vermontrepublic.org

Vermont Republic's website.

Vermont Commons
www.vtcommons.org

Blog/journal dedicated to the proposition that Vermonters should peaceably secede from the United States and govern themselves as an independent republic once again.

SELF-EMPLOYMENT IDEAS

We can't make any guarantees, but these are good possibilities. And they should get you thinking in the right direction about other likely businesses.

Barter Network

When people don't have money, they barter. They've always got stuff—and skills—that they can exchange. (See page 125.)

Beer and Wine Making

No matter how hard times get, people will still want beer and wine. If you can turn them out at home, you've got an endless supply of barter material.

Bicycle Sales / Repairs

The first thing you should do right now is run out and buy as many used bicycles as you can. Used bike sales and repairs should do very well. You could even add motorcycles and scooters, because of the good gas mileage they get.

Boarding House

If your home is big enough, or can be made big enough, open a boarding house. Offer rooms on a weekly or monthly basis, and include as many meals a day as you feel you're up to. You should probably at least include breakfast and likely even dinner, giving your guests a choice as to which plan they prefer.

Car Repair

While this is no time to be in the *new* car business, car repair should be fine as people try to keep their existing vehicles running as long as they can—and as long as they can afford the gas. Consider basic car repair/maintenance where you go to the customer's home to do the service. You might even have luck with ongoing maintenance contracts. You could also give car repair classes.

Cheap Luxuries

Even in hard times—actually *particularly* in hard times— people will want to spoil themselves now and then. But the luxury has to be cheap. It might be a special chocolate chip cookie, or delicious homemade candy. Or maybe it's very special handmade gift wrapping paper. Or fragrant handmade soap or a hand-carved wooden toy.

Computer Repair

Computer repair and maintenance should do exceptionally well. People will no longer be running out to buy the latest upgrade (computer, monitor, printer, etc.) but will be more interested in keeping what they have working properly.

Consignment Shop

Consignment shops don't have to be limited to clothing. Garden and automotive tools, household appliances, furniture.

Delivery Service

It makes much more sense to pay a slight fee to have a store deliver a purchase to you than to get in your car and make a round trip to do it yourself. Smart store owners will offer *free* delivery in order to attract customers. On a larger scale, companies like UPS and FedEx should do well. On a small scale, why not start your own local service? You might even consider moving people around as well as packages.

Energy Consultant

There is, and will be even more, a demand for information on how people can cut their utility bills. Armed with the information in this book, and your own research, you could offer local classes on energy conservation (and thus money conservation), and even go into homes giving people specific advice on how to save money by using less electricity and gas.

Entertainment

People want to be entertained, particularly when times are tough. Consider bringing together talented people to offer live entertainment in neighborhood or community locations. You probably won't get rich but you'll have a good time, and you and the entertainers will end up with more money than when you started.

Errand Service

On a more personal level than a delivery service, errand services can combine your needs with similar needs of others, to provide services cheaper than you could do them yourself. This might be pickup and delivery, but could also include banking, taking children to after-school classes and sports or pets to the veterinarian, and a variety of other activities.

Fads

It would be nice to invent, and get rich off, the next fad. But you can also stay in touch with the culture, and offer low-cost ways for your community to enjoy those fads.

Farming

You don't have to have a huge farm to have a healthy farming business. An acre of land and a lot of hard work will produce what you need for a small-scale business. Consider specialty items such as mushrooms, herbs, or sprouts. Because of the many uses of hemp, that will be a great crop once it's legal—or at least possible without enforcement—to grow (See page 212).

Herbal Treatments

You can grow, or forage for, herbs and use them to make healing concoctions, syrups, salves, teas, and a variety of other healthy items (See page 189). You could also use flowers for Aromatherapy (See page 186).

Home Repair

People will have to do what they can to keep everything in their home in working order. If you have carpentry, plumbing, electrical, or other practical skills, you can be a big help to those people. You might even have success at teaching those skills to others.

Instruction

You name it, someone will be interested in it. Whether it's using tools, playing music, sewing or storytelling. The lessons likely to do best will be those focused on basic essentials, saving money and protecting health, such as gardening, food canning, inexpensive home cooking, yoga and other health exercises, meditation and relaxation, herb foraging and use, and pet care.

Instructional Video Library

Considering setting up an instructional video library; videos
and DVDs that teach people various skills, whether they're
home or car repair, sewing, musical, language or any of
hundreds of other subjects.

Raising Chickens and Rabbits

Lots of people might start their own gardens, but very few will
raise their own meat. You can take advantage of that by
raising chickens and rabbits—and other critters if you've got
room (See page 93).

Rental Library

Libraries aren't just for videos and books. Consider offering specialty kitchenware, car and woodworking tools, games, toys, household repair tools, gardening equipment and literally anything else you can think of. However, do this for your community, not your neighborhood. In your neighborhood, you should simply be sharing.

Seamstress / Tailor / Clothing Alterations

In hard times, people can't afford to simply buy new clothes to replace slightly worn clothing. Darning socks will be back again. Sewing and knitting skills are no longer common, nor are sewing machines. Anyone with these skills will be much in demand.

Knitting Help
www.knittinghelp.com
More than 150 free online knitting instruction videos.

Sleepwear

People will be saving money by not using their gas or electric heating at night. Help them stay warm by making and selling sleepwear, especially such items as head and foot coverings. Nightcaps and night socks will keep them warm by preventing loss of body heat.

Small Appliance / Electronics Repair

Until now it's been cheaper to throw away a broken radio, telephone, blender, microwave, coffee maker or similar appliance than to have it repaired. That time is over. With the end of cheap imports, and loss of income, people will no longer be able to afford new items. They will either have to do without, or find someone with the skills to fix them.

Soap Making

Everybody needs soap. While there might not be a time when

you can't find soap to buy, you'll save money, have some fun, and end up with an excellent barter item by making your own soap.

Soap Recipes
www.waltonfeed.com/old/soaphome.html
www.soapnaturally.org

Skin Care Recipes
www.soapnaturally.org/recindex.html#skin

Toys and Games

There was a time when toys were not mass-produced plastic things. They were carefully crafted from wood, cloth and other natural materials, were treasured by the children who received them, and were passed down from generation to generation. Create some yourself, and you'll have customers.

SOCIETIES, EMPIRES AND CYCLES

"The farther backward you can look, the farther forward you are likely to see." *Winston Churchill*

"Look back over the past, with its changing empires that rose and fell, and you can foresee the future, too." *Marcus Aurelius*

Societies collapse for reasons, and social scientists and writers see commonalities in those reasons. Some see environmental causes, others believe that the causes are inherent defects in the society that are ignored in the short run but that bring down the society in the long run. Still others believe that collapse is simply a natural result of the cyclical nature of history.

While there are differing theories as to *why* societies come to an end, there is total agreement that no society has yet avoided its inevitable destiny.

There is no reason to believe that our society will be the exception. But perhaps we can do some things to stall that end.

CYCLES

Is everything new and never before seen? Are our actions the only factors affecting the course of history?

Throughout the history of humanity many people have studied trends and cycles and concluded that there are natural forces much greater than we that affect the destinies of our societies and our lives.

Perhaps any economic and societal crash we experience has less to do with Peak Oil, budget deficits and housing bubbles than it does with the cyclical ebb and flow of history.

Cycles E-books [free e-books]
www.techsignal.com/fsc_ebooks.htm

The Case for Cycles, Definitions of Cycles, Cycles and War, and Cycles - The Mysterious Forces That Trigger Events.

Cycles Research Institute
www.cyclesresearchinstitute.org

For the interdisciplinary study of cycles.

Foundation for the Study of Cycles
www.foundationforthestudyofcycles.org

Purpose is to discover the causes and conditions for cyclic and rhythmic behaviors.

The Fourth Turning [book]
Author: William Strauss and Neil Howe

The past and future of Western history.

The Great Cycle [book]
Author: Dick Stoken

Predicting and profiting from crowd behavior, the
Kondratieff Wave, and long-term cycles.

Kondratieff Theory
www.kwaves.com/kond_overview.htm

Includes biography of Nikolai Kondratieff.

Kondratieff Wave
www.angelfire.com/or/truthfinder/index22.html

A 54-year (average) economic, social and cultural cycle.

Kondratieff Waves in the U.S.
www.thelongwaveanalyst.ca/cycle.html

From 1789 – 2003.

The Long Wave and War
www.timesizing.com/1kondrat.htm

The influence of cycles on war.

Thoughts on a Second Great Depression
www.gold-eagle.com/editorials_05/nystrom030905.html

One person's commentary on cycles.

The Asian Slump - Will It Become a World Depression?
www.ru.org/81batra.html

Dated essay by economics professor Ravi Batra, but the
Law of Social Cycles that he presents is very interesting.
Batra has been described as "having predicted 10 out of the
last 3 recessions" (he was wrong on the timing of the one
discussed here, too).

COLLAPSE OF CONTEMPORARY SOCIETIES

Societal collapse isn't just in the distant past. It's all around us. Perhaps in our present, definitely in our future. We can learn much from societies that have recently had such collapses, such as Argentina, Cuba (see more in this book at page 205), and the Soviet Union.

> Contemporary Collapses
> www.beyondpeak.html#contempcollapse
>
> How Argentina, Cuba and the Soviet Union collapsed, and how the people responded. Includes articles on the American Empire.

POSSIBLE SCENARIOS

Nobody knows for sure how a Peak Oil-instigated economic collapse might play out, but there are some pretty interesting guesses.

> Possible Peak Oil Scenarios
> www.drydipstick.com/peakoil-scenarios.html
> www.beyondpeak.com/scenarios/winners.html
>
> Various scenarios available on the Internet as well as the winners of the First Annual Beyond Peak Scenario Contest.
>
> Potted History
> www.darkage.fsnet.co.uk/PottedHistories.htm
>
> A brief history of every civilization and empire that rose —and fell (That's every one of them, so far.)
>
> Why Societies Collapse
> www.abc.net.au/rn/talks/bbing/stories/s707591.htm
>
> Transcript of a talk by UCLA professor Jared Diamond, winner of a 1997 Pulitzer Prize for his book Guns, Germs and Steel.

Collapse [book]
Author: Jared Diamond

How Societies Choose to Fail or Succeed, by the author of *Guns, Germs and Steel*.

The Collapse of Complex Societies [book]
Author: Joseph Tainter

A study of why societies have collapsed throughout history.

The Coming Dark Age [e-book]
Author: Roberto Vacca
www.printandread.com

On the risks of unmanageability of large technological systems and the tragic consequences it could have for the world. [e-book]

The Death of Megalopolis [e-book]
Author: Roberto Vacca
www.printandread.com

A futuristic doom novel.

Dark Ages America [book]
Author: Morris Berman

The final phase of empire.

PEAK OIL RESOURCES

Peak Oil Websites

Dry Dipstick
www.drydipstick.com

Metadirectory of the best Peak Oil websites.

Beyond Peak
www.beyondpeak.com

Prepare for Peak Oil and economic collapse.

Peak Oil Prep
www.peakoilprep.com

All the books, DVDs and websites mentioned in this book with direct links to their source.

Energy Bulletin
www.energybulletin.net

Latest Peak Oil news.

The Oil Drum
www.theoildrum.com

Intelligent, informed discussion on energy depletion.

Oil Depletion Protocol
www.oildepletionprotocol.com

Home of the effort to have the protocol adopted.

Community Solution
www.communitysolution.org

Helping small, local communities respond to Peak Oil.

Post Carbon Institute
www.postcarbon.org

Helping people and communities prepare for Peak Oil.

Peak Oil Books

Powerdown
Author: Richard Heinberg

Options and actions for a post-carbon world.

The Party's Over
Author: Richard Heinberg

Oil, war and the fate of industrial societies.

High Noon for Natural Gas
Author: Julian Darley

The other fossil fuel that's about to peak.

Twilight in the Desert
Author: Matthew Simmons

The coming Saudi oil shock and the world economy.

The Long Emergency
Author: James Howard Kunstler

Surviving the end of the oil age, climate change, and other converging catastrophes of the twenty-first century.

The Coming Oil Crisis
Author: Colin J. Campbell

By the founder of the Association for the Study of Peak Oil.

Beyond Oil
Author: Kenneth S. Deffeyes

The view from Hubbert's Peak.

Crossing the Rubicon
Author: Michael Ruppert

The decline of the American empire at the end of the age of oil.

Peak Oil Videos (DVD)

The End of Suburbia
www.endofsuburbia.com

Oil depletion and the collapse of the American Dream.

Escape from Suburbia
www.escapefromsuburbia.com

Beyond the American Dream.

Peak Oil Websites

Association for the Study of Peak Oil
www.peakoil.net
International network of scientists.

Energy Bulletin
www.energybulletin.net
Energy and Peak Oil news.

Global Public Media
www.globalpublicmedia.com
Audio and video on the post-Peak Oil age.

Peak Oil in the News
www.peakoilinthenews.com
Chronicling Peak Oil as it happens.

Peak Oil
www.peakoil.com
News and message boards.

PowerSwitch
www.powerswitch.org.uk
Peak Oil in the U.K.

Life After the Oil Crash
www.lifeaftertheoilcrash.net
Popular site with news, articles and forums.

More Peak Oil websites
www.drydipstick.com/peakoil-websites.html

Index

Peak Oil Prep

Printed in the United States
90279LV00006B/26/A